Is Theology a Science?

Studies in Systematic Theology

VOLUME 7

Is Theology a Science?

The Nature of the Scientific Enterprise in the Scientific Theology of Thomas Forsyth Torrance and the Anarchic Epistemology of Paul Feyerabend

By

David Munchin

BRILL

LEIDEN • BOSTON
2011

This book is printed on acid-free paper.

Library of Congress Cataloging-in-Publication Data

Munchin, David.
Is theology a science? : the nature of the scientific enterprise in the scientific theology of Thomas Forsyth Torrance and the anarchic epistemology of Paul Feyerabend / by David Munchin.
p. cm. -- (Studies in systematic theology ; v. 7)
Includes bibliographical references and (p.) index.
ISBN 978-90-04-19459-5 (hardback : alk. paper) 1. Religion and science. 2. Torrance, Thomas F. (Thomas Forsyth), 1913-2007. 3. Theology--Methodology. 4. Feyerabend, Paul, 1924-1994. 5. Science--Philosophy. I. Title. II. Series.

BL240.3.M85 2011
201'.65--dc22

2011008582

ISSN 1876-1518
ISBN 978 90 04 19459 5

For Ysmena

CONTENTS

PREFACE

This book began life as doctoral thesis, and therefore is primarily a work of academic scholarship. I am not however a professional academic but a parish priest in the Anglican Church, for whom at one level, work such as this is a diversion, therapy and intellectual stimulant. However at a more significant level, it is also an attempt – and hopefully this is not too pretentious a claim – to think hard about the intellectual challenges and resources of the Christian gospel, as part of a ministry that is rooted in the day to day joys, trials and muddle of people's lives. It is fuelled by a belief that somewhat contrary to present trends, any too permanent or deep divergence between Church and University, will in the long term, only do damage to the standing of the gospel, as well as marking a break in the tradition that numbered thinkers and doctors among the saints. That is not to say that the church will not have profound and searching questions to ask the University – Barth for instance, would certainly have reserved that right to the church – but it must do so as critical friend, and not as a refuge of the disenfranchised, exempting itself from the call to an honest and genuine dialogue. This is not simply because the church has much to gain pragmatically from allying itself with scholarship, but for the rather more important reason that 'truth matters', even if, as with Feyerabend, we have to wrestle the concept of truth from ironic quote marks. Truth matters not just in the University, but in all human lives – *pace* Feyerabend, I believe that the search for truth and human flourishing are ultimately coincident. The forces ranged against such a search are in the ascendant – scholars are not wholly innocent, but more to blame are the media, advertisers and politicians, who have often found the truth uncomfortable and inconvenient. However, neither has the church been innocent, as pressure grows to 'package' a 'needs-and-wishes-answering' faith in dangerously simplistic and un-nuanced form.

Contemporary culture's preference for simplistic and antithetical disjunctions – fact and value, fact and opinion, faith and reason – is deeply pervasive and damaging. This book is concerned with a further dichotomy. "To assert that a person, or his idea, is 'unscientific' is the severest charge."[1] Faith and science have become antithetical terms, the former often lazily assumed to be simply a series of pragmatic psychological projections. Such a dichotomy is dry and potent fuel to the gospel's modern day cultured despisers, grown fat upon the legacy of centuries of supposed conflict between science and religion.[2] However the conflict hypothesis ignores the work of philosophers of science, who note that 'scientific' is a contested notion. What counts as 'scientific' or often 'scientific evidence' is itself conditioned by political, philosophical and cultural factors. Examination of the notion of the 'scientific' reveals that our lazy dichotomy will not do – faith is not opposed to science, and by implication scholarship, warranted knowledge and ultimately truth. My way into that question is through Torrance and Feyerabend, and at the outset I offer my reasons for selecting that pairing – but that is not to say that my conclusions are thus limited – we can ask the question more generally or to other particular thinkers. Faith has everything to lose by being drawn into a discussion which assumes the conflict hypothesis as fact – only by questioning the fundamental epistemic assumptions implicit within that hypothesis can progress be made. Moreover in facing those questions we stand on the shoulders of giants – Paul, Athanasius, Augustine, Aquinas, Calvin, Barth.

When Thomas Forsyth Torrance was an Army Chaplain, during the attack on the San Martino-Sogliano ridge, he, under fire, found himself comforting a soldier dying of his wounds. The question which that soldier asked Torrance as he died, stayed with Torrance for the rest of his life: 'Padre, is God really like Jesus?' It is not too exaggerated a claim that Torrance's distinguished career as a scholar was dedicated to answering that question.[3] It brings to mind someone in a parish discussion group, who whenever the question came up would say: "Jesus, yes, he's lovely and kind; but God, I don't know, you've got to be a bit nervous of him." My poor efforts to explain the dual nature of Christ and the doctrine of the Trinity did nothing to lessen the anxious repetition. The theological questions which scholars address are not irrelevant self-referential esoterica, but impinge upon vital questions of human existence, as the above questions from the battlefield and discussion group testify.

---

[1] Gelwick, 1977, p. 14.
[2] Luoma, 2002, p. 3.
[3] McGrath, 1999, p. 73f.

ACKNOWLEDGEMENTS

This book has as its genesis, a doctoral thesis completed at Heythrop College, London University, in 2009, and so many thanks must first of all go to my supervisor Prof. Gwen Griffith-Dickson, whose intelligence, compassion, humour and hospitality saw me through the highs and lows of that process. I would also like to thank the other staff and students for help and support. As a parish priest, I must also thank my colleagues and the people of Hatfield Hyde and Welwyn who have given the time for me to work on this book, and tolerated a pastor whose mind was often full of things with only a tangential relation to parish life. Thanks must also therefore go to the Diocese of St Albans, and in particular to Bishops Christopher Herbert and Alan Smith, and Canon Dr Dennis Stamps, who not only gave me time for this work, but also granted and funded a sabbatical, and provided active support and encouragement. Thanks to my proof reader Jean Reynolds. I was much tempted to dedicate this book to the labradors – Cella, Tom, Belle, and George – who twice a day everyday made me stop, walk and think – a good discipline for scholars and pastors alike. Finally however there was only one candidate for dedicatee, Ysmena, whose love, support and scholarship was and is a foil to my often tired and ill-tempered ways.

NOTE TO THE READER

Several of Feyerabend's books were published in the same years, so for ease of reference, in footnotes I have for Feyerabend used the format of 'Author: initials of title: page no.' whereas elsewhere I have used 'Author: date: page no.', though for Barth's *Church Dogmatics*, I have used 'Barth, CD Vol no., page no.'

Quotations from the Bible are from the New Revised Standard Version.

CHAPTER ONE

INTRODUCTION: CONTEXT AND HISTORY

Since its very beginnings theologians have sought to elucidate the relationship between the Christian gospel and the secular learning of its host culture. Both the cultures, and the degree of interaction with or hostility to them, have varied greatly. Judaism, through thinkers such as Philo, had already sought to engage positively with the Hellenistic philosophy which marked the gold standard of pagan intellectual life, and in particular bequeathed to Christianity the problem of how learning derived from experience of the material world could be reconciled with a faith based on truths revealed by God. From the New Testament onwards – Paul debating in the Areopagus (Acts 17) being the obvious example – through Fathers such as Justin Martyr and Tertullian, to the present day, the question, posed in every conceivable human situation, has found a variety of solutions: syncretic, synthetic, dialectic and antithetic.

In the modern era the rising importance of empirical natural science has meant that it has come to be held in the same esteem as were the writings of the great Greek philosophers in antiquity. Thus a scientistic identification of 'modern thought' as more or less synonymous with modern natural scientific thought in the popular imagination. Such an identification has therefore brought renewed and increased pressure upon Christian theology to define how its epistemological commitments stand in relation to the truths discovered and expounded by natural science – especially when, even if mistakenly, such truths have been seen at face value to be incompatible with Christian teaching. Again Christian theology has reacted in a number of ways – from a simple fundamentalist reassertion of the primacy of 'God-given' revelation over mere human learning, to a radical confinement of religious truth to the symbolic expression of emotive hopes, ideals and aspirations. One of the points of this dialogue is try to establish where a present day theologian might best position a mediating standpoint between these extremes.

The pre-eminence of 'science', its function as a hallmark of epistemic rigour, has also given new impetus to the argument over whether theology itself is a science and, if it is, what sort of science. As we shall see this is a question with a history. An important point in that history is reached with the assertion with which Barth opens his *Church Dogmatics*: 'Dogmatics is a science.'[1] Barth's own affirmation, as I shall show, is predicated upon repudiating antecedent grounds for an affirmative within the liberal Protestant thought of the preceding two centuries. Although it was this liberal Protestantism which had nurtured Barth, he believed that its petition for scientific status had only been granted at the cost of theology forfeiting its own integrity. Rethinking the grounds for this petition provided him with the necessary purchase for a new approach.

This relation between natural philosophy and Christian theology is by its nature perennially in need of restatement because here we are dealing with two dynamic disciplines, each with their own trajectories of development and evolution. There can be no normative moment, such as was arguably achieved at Chalcedon or Nicea, at which Christian theology can once and for all state its assessment of and relation to natural science, because, particularly in today's context, that could only amount to passing judgment on the natural science of yesterday. Theology must take account of and account for, the shifting nature of scientific knowledge, as well as respond to pressures, discoveries and fashions within its own discipline. Indeed the fact that natural science is constantly changing, sometimes in a way which repudiates previously strongly held theories, should serve as a warning to those who criticize theology for not having an easily identifiable and unshifting core paradigm, by demonstrating that change and revision is always a product of the scientific method.

Not only does the content of natural science change, but particularly in the past century, there has been an increased interest in the distinct discipline of the philosophy of science. Such a discipline deals with questions concerning the nature and the boundaries of the scientific enterprise: what are the defining characteristics of a science? How has science functioned (descriptive)? How can we tell good science from bad (prescriptive)? Also, particularly since the controversy surrounding logical positivism, it seeks to evaluate the relative epistemic value of the findings of the various sciences and non-sciences, however those taxonomies are drawn up. Theologians have therefore found it fruitful

[1] Barth, 1949, opening words, also in CD I/1, p. 1f.

to use the categories of the philosophy of science to answer questions relating to the scientific status of their own discipline, and its relation to those disciplines considered exemplarily scientific. This raises both a semantic problem – what activities fall under the scope of the word 'science'? – and a more epistemically substantive question about the nature of the scientific enterprise.

So for instance Nancey Murphy (1990) seeks to evaluate Pannenberg's project (and those of a set of modernist catholic theologians) utilising Lakatos' category of 'research programme', and judges the former to qualify as a 'progressive' example of the latter. Incidentally her work offers therefore both a precedent for this book, and a spur, with her remark "it appears that no one has tried to show the relevance of Feyerabend's philosophy for theological work."[2] Pannenberg[3] sought to accommodate theology to the scientific criteria of Heinrich Scholz, an attempt which Barth had so forcefully rejected – we shall have cause to return to the details of the Barth – Scholz encounter.

Thomas Kuhn's conceptions of scientific paradigms shifts, normal and revolutionary science, have found extensive employment beyond the confines of natural science, notably in the social sciences. Küng and Tracy edited (1989) a set of papers from an important symposium (*Paradigm Change in Theology*) which took as its starting point these Kuhnian notions. Most importantly for this book Thomas Forsyth Torrance made extensive, though not uncritical or uncontroversial,[4] use of the ideas of Michael Polanyi in his project of theological science.

---

[2] Murphy, 1990, p. 84; also Griffith-Dickson [Feyerabend] "might be a starting point, but rarely is", 2005, p. 40.

[3] Holwerda in Plantinga & Wolterstorff (eds.), 1983, p. 281.

[4] Polanyi's experiences no doubt led him to an outlook that was deeply suspicious of fanaticism and enthusiasm, and his faith was characterized by a moderate and temperate outlook, particularly in his last book *Meaning*, written with Prosch, which underlines the symbolic and mythical qualities of religious language. Indeed towards the end of his life, his religious views shifted again toward the 'cosmic theologian' Mircea Eliade. According to Torrance, *Meaning* marked a change of direction for Polanyi. However Weightman insists that *Meaning* is in fact an organic development of Polanyi's views, albeit one of which Torrance did not approve (Weightman, 1995, p. 114). According to Weightman, Polanyi's reticence when it comes to theological specifics allows his philosophical ideas to be theologically 'completed' in a variety of ways. Indeed, according to Weightman, the more 'natural' completion is towards Tillich rather than Torrance (Weightman, 1994, p. 77). Also according to Weightman, Polanyi's approach to the birth, death and resurrection of Christ, is much closer to that of Tolstoy than of Torrance.

Torrance is perhaps chiefly known as pupil, translator, interpreter and champion of the theology of Karl Barth in the English speaking world.[5] His next most influential contribution to theology concerned the relation of Christian theology to the natural sciences, for which he won the Templeton prize in 1978. Torrance's interest in this field spans almost half a century, his landmark book (*Theological Science*) being published in 1969. Ironically it is with regard to this relation that the distance between Barth and Torrance was at its greatest. However Torrance did not himself admit to any particular un-Barthian turn of thought, arguing instead that it was changes in modern natural science

---

Weightman's contention that Polanyi 'lends' himself to Tillich-like rather than Torrance-like completion is a difficult one to weigh up. In *Meaning*, which Weightman refers to, Polanyi specifically rejects any programme of demythologising (Polanyi & Prosch, 1977, p. 158). Scott also describes Tillich as the theologian closest to Polanyi, though she immediately refers to their differences over the relation between theology and natural science, and later cites Torrance as being closer to Polanyi on this particular point (Scott, 1995, p. 182f). Polanyi attempts to broaden an epistemology confined by the ideology of scientism by illuminating the personal and tacit elements within all science. Being primarily a student of the philosophy and history of science, whilst he his remarks are suggestive for theology, they are not prescriptions:

> Christianity is a progressive enterprise. Our vastly enlarged perspectives of knowledge should open up fresh vistas of religious faith. The Bible and the Pauline doctrine in particular, may still be pregnant with unexpected lessons; and the greater precision and more conscious flexibility of modern thought, shown by the new physics and the logico-philosophic movements of our age, may presently engender conceptual reforms which will renew and clarify, on the grounds of modern extra-religious experience, man's relation to God. An era of great religious discoveries may lie before us (Polanyi, 1998, p. 285).

That 'opening up of fresh vistas' no doubt includes Tillich and Torrance, but its openness is genuine enough that Wightman's attempts to over-specify are ultimately underdetermined. Torrance's view is that such openness is merely necessary but not sufficient for his Barthian style of theology – moreover that *'non-sufficiency' is of itself a vital part of such theology*. If Polanyi (or Einstein or Gödel or quantum mechanics) did of necessity *imply* a Barthian position, then it would be implied upon extra-theological grounds, and the floodgates of natural theology would be open – only by demonstrating that Barthian theology is one option amongst several legitimate 'Polanyi-completers', can that theology stay true to itself. If Torrance sometimes oversteps the mark, and makes too strong a connection between natural science and theology, then that enthusiasm must be curbed.

However what is not in doubt is the importance for Torrance of Polanyi's work, and their mutual regard for one another as scholars and persons: Torrance edited *'Belief in Science and in Christian Life: The relevance of Michael Polanyi's thought for Christian faith and life'*; contributed a second preface to Scott's biography of Polanyi; published numerous papers and books which made extensive reference to his works; and was Polanyi's literary executor.

[5] McGrath, 1999, p. 112.

(most importantly relativity theory) and less positivistic approaches in the philosophy of science (Polanyi and Popper) which accorded Barthian theology its necessary epistemic workspace. According to Torrance, twentieth century developments in natural science had profound resonances with a Barthian approach which rejected dualisms, they themselves being artefacts of ancient philosophies. Those developments also undermine the unyielding positivism which entailed a sharp demarcation between observation and theory, and hence observer and observed. What Torrance sought was a unitary theory of knowledge,[6] which would grant to both theology and other scientific disciplines their integrity and independence, but which would also underpin a strongly realist potential for truth telling within an interconnected and holistic conception of reality and knowledge.

There are several possibilities for assessing Torrance's project. One possibility would be a straightforward exposition and critical appraisal – on the grounds of the internal coherence of the project, and its relation to alternative points of view from the worlds of theology and the philosophy of science. Such attempts have been made, notably by McGrath in his *Intellectual Biography* of Torrance in 1999, and also in a trio of volumes written or edited by Elma Colyer. Elements of such an approach are certainly present in this book, but apart from already having been done, another disadvantage is that it lacks critical focus: in assessing Torrance with reference to the *whole* of theology *and* the philosophy of science, we set ourselves an ambitious task which risks simplifying or generalising results from these two complex fields.

Instead, as with Murphy on Pannenberg, I set myself the task of submitting Torrance's approach to a challenge test by a critical friend – Paul Feyerabend. As we shall see Torrance's conception of science hinges on a revised understanding of objectivity, Feyerabend seeks to show that his dissembling attack on foundationalism undermines such an understanding, along with other categories integral to the project, namely rationality, realism, and indeed truth.

Paul Feyerabend is also perhaps best known as a pupil – in his case of Karl Popper. However unlike Torrance and Barth he was ultimately no champion of Popper's views. Rather as he grew away from Popper's influence, he began to regard Popper as embodying everything within the philosophy of science which was self-satisfied and self-serving.

[6] McGrath, 2001, Vol. 1, p. 21; Colyer introduces the label 'holism', 2001a, p. 16.

Feyerabend's reputation became that of an 'anti'-philosopher of science – an anarchic, purifying and therapeutic figure who sought to open up the windows of the stuffy scientific laboratory and allow the whole gamut of human truth-seeking traditions epistemic breathing space.

There is therefore a fundamental asymmetry to the project, for the anarchist has no programme to compare, no alternative project to promote. Furthermore not being a theologian, Feyerabend's challenge test will be concerned with questions of epistemology and scientific method, not theology – though as we shall see, he is not shy in commenting on matters theological. *Rather Feyerabend the anarchist is trying to undermine Torrance's project by denying coherency and substance to the predicate 'scientific'. Not only that, but he will initiate that process with a sustained attack on positivistic conceptions of science, a hostility which has much in common with Torrance – so that by diverging towards more radical conclusions, he will be turning Torrance's critique back on Torrance himself.* This is not simply a comparative study of two *differing* conceptions of 'the scientific'; the dynamic of the encounter by its nature is lop-sided, so that the question is whether Feyerabend can successfully deconstruct Torrance's scientific predicate, with no expectation that he will offer us an alternative.

There is no doubt that after his advocating of 'epistemological anarchy', Feyerabend's reputation within the community of philosophers of science suffered, though ironically his engagement with wider social and educational questions offered him a new and appreciative audience beyond that community. One must therefore state clearly how Feyerabend stands with regard to the community of philosophers of science. Certainly any philosopher of science who was labelled *The worst enemy of science*[7] by friends and critics alike, cannot be understood as in any way 'representative' of that community. Indeed his anarchic and deconstructive approach led the critics to write him off as a sort of heckler shouting epistemic obscenities from the sidelines. However particularly in chapter five, I seek to show that there is a serious, if somewhat disjointed core of ideas which define his epistemological anarchy, and offer a real and serious challenge to more conservative philosophies of science. Feyerabend deserves to be taken seriously – or at least as seriously as he took himself. Furthermore as

[7] Followed by a '?' the title of a volume of essays written in his memory – Preston, Munevar and Lamb (eds.), 2000.

we noted, whilst his stock has fallen within the philosophy of science community, many of his ideas, such as incommensurability,[8] have yielded high value by their recycling in other communities academic (e.g. J. L. Lyotard) or otherwise, sheltering under the post-modern umbrella,[9] which in turn have influenced theological developments in the past thirty years.

* * *

What we therefore have is a constructed and critical dialogue, in which Torrance expounds his project – theological science (which includes the notion of the 'scientific' as a revised 'objectivity'). It is then submitted to Feyerabend's challenge (essentially an attack on foundationalism, which he pursues through to a deconstruction of the 'scientific', including 'objectivity'), and Torrance replies to that challenge (the revised 'objectivity' is immune to such deconstruction). It is constructed in that there was never any actual dialogue, verbal or written, between them. It is critical in that a third voice, my own, will be heard assessing and criticising their contributions, and drawing conclusions.

My conclusion will be that Feyerabend's deconstructing radicalism is fuelled by a series of false dichotomies (or at least a hardening of continua into dichotomies),[10] having often, in an historically dubious manner, floored a 'straw man' of his own invention, a history in which traditions only collide and never meet. His attack on 'foundationalism' ultimately does not entail the more radical conclusions which he supposes, and the revised conceptions of rationality, realism, and in particular objectivity, which inform Torrance's conception of science, emerge from Feyerabend's challenge more or less in tact – certainly Feyerabend's prescription that we should forego such conceptions altogether is a failure. This is despite Feyerabend's formidable and often entertaining rhetorical barrage. Torrance I believe survives, not without sustaining some shrapnel wounds, by positing a more nuanced approach, admitting of a continuum of viewpoints rather than Feyerabend's bi-polarism.

---

[8] E.g. J. M. Foster, *Valuing nature? Ethics, economics, and the environment*, Routledge, 1997; A. Pennycock, *Incommensurable discourses*, Applied Linguistics 15, Oxford, p. 115–138; H. Aldrich in *Rethinking organisations: New directions in organisation theory and analysis*, M. Reed & M. Hughes, Sage, London, 1992.

[9] Preston, 1998, p. 7.

[10] Putnam, 1987, p. 27, cites a conversation with Chomsky in which he bemoaned this tendency in philosophers.

What then are the nature and seriousness of those wounds? In particular I posit the view that if we deny Torrance those epistemological supports which have a decidedly theological flavour, then the situation begins to look a lot closer to the anarchic one which Feyerabend envisages. However this must not be understood as a rather banal conclusion – after all Torrance is a theologian and a Christian, Feyerabend is a philosopher of science with no particular religious commitments – as though Torrance is Feyerabend with Christianity tacked on the end, or dressed up in Christian clothing. Rather, theological ideas, chiefly creation and incarnation, permeate deeply into the whole of Torrance's epistemology – *his epistemology is best characterised as itself Christian.* Torrance does not offer Christian theology as *one* way out of certain philosophical quagmires, as though secular philosophy might offer alternative routes, but Christian theology, for good *theological reasons*, offers the *only* way out. God is not simply an object which might be missed off an audit of the universe, but rather as its creator, redeemer and sustainer, the universe is ultimately, for Torrance, incomprehensible without him. Finally we conclude that for Torrance truth and freedom are united under God – if we aim for one we also converge towards the other. For Feyerabend they are *opposed* – aiming for one (truth), can only distract us from our pursuit of the other (freedom).

* * *

There is however a semantic issue to address immediately. The English word 'science', as the OED notes, has in general use sharpened its meaning to 'natural science',[11] and where other uses are intended, this is indicated by prefixing other terms, e.g. 'social', 'human', etc., lending an ambiguity to the terms 'scientific' and 'the sciences'. In this book I shall use the term in both its wider and narrower senses, only adding the prefixes 'natural', 'theological', 'human' etc. where the context is ambiguous. Although both Torrance and Feyerabend wrote principally in English, it is worth noting that German is Feyerabend's first language, and that he came to the English speaking world (of Bristol University) only in 1955.[12] Also Torrance's theological inheritance, including being offered Barth's chair at German speaking Basel, is mainly German. Furthermore the question of the scientific status of theology in previous

[11] OED marks shift in 1867 edition.

[12] In Preston's list of Feyerabend's published works there is only one German paper, from 1960.

decades has its *locus classicus* in Barth's discussion with Scholz, and subsequently his employment of the formula 'Dogmatics is a science' at the opening of the *Church Dogmatics*. Barth's sentence is translated (by G. T. Thomson) from the German, a language in which the word science "sounds even more impressive – *Die Wissenschaft*."[13] Regardless of Feyerabend's aesthetic judgement, words in one language may not be isomorphically corresponded to words in other languages, and this is true of the words 'science' and "that untranslatable word *Wissenschaft*."[14] *Wissenschaft* (and its adjectival equivalent *wissenschaftlich*) does indeed have a wider or at least different scope from the word 'science'. Its standard translations include 'science', 'scholarly' and 'academic'.[15] Moreover mirroring the shift in English usage, the introduction of the term *Geisteswissenschaft* upon the translation of John Stuart Mill's *Logic*, and taken up by amongst others Dilthey, marks a similar attempt to clarify a parallel ambiguity. As we shall note Dilthey is here not simply attempting to demarcate natural science from the human sciences semantically but methodologically – hermeneutical and empirical methods being respectively and exclusively the proper methods of the human and physical sciences. This hard distinction is rejected in the universal or philosophical hermeneutics of Gadamer. So this difficulty is not simply a matter of translation, demarcation or typology, but the sense and reference of these terms are shaped by the epistemological and methodological issues bearing upon our discussion.

Contextual background on the scientific status of theology until Torrance

According to Pannenberg, the question of theology's scientific status rose first to prominence in the thirteenth century. Within a century not only had theology become designated as a science, but as the highest science. He goes on to points out that "previously, [before C13th] Christian teaching, under the influence of Augustine, had been regarded as *sapientia*, in sharp distinction from *scientia*."[16]

[13] Feyerabend, AM, p. 246.

[14] Frei, 1992, p. 97.

[15] For instance, in Oxford and Collins German-English dictionaries, though the two directions are not symmetric.

[16] Pannenberg, 1976, p. 7f – This distinction is founded in Augustine's neoplatonism, and was, as Polanyi and Torrance both note, to have important historical implications for the development of both theology and natural science.

Pannenberg expounds *scientia* in its Aristotelian sense as conclusions syllogistically derived from self-evident principles. Pannenberg goes on[17] to trace further discussion of this distinction to mid C12th Chartres, when Nicholas of Amiens in his *Ars Catholicae Fidei,* used Euclidean geometry as a model for the development of his philosophical theology, though the axioms that Nicholas used were by and large considered to be articles of faith, and not the self evident ones of Aristotle's *Posterior Analytics.* Aquinas categorized theology as a derived science, again in the sense of the *Posterior Analytics* (i.e. analogous to optics and harmonics, which were derived from or subordinate to geometry and arithmetic respectively), because its axioms were not self-evident but derived from elsewhere.[18]

Duns Scotus, *pace* Aquinas, but again using the Aristotelian taxonomy, categorized theology as a practical science, because it proceeds *towards* a known goal, rather than *from* axioms. Wide acceptance of the validity of Scotus' objections to Aquinas meant that not until the rise of neo-scholasticism was Aquinas' designation to regain importance. One can discern here the seeds of debates concerning the relative merits of theoretical and practical disciplines, and how practical a purpose theology should serve, which have a sure contemporary resonance.

Luther followed Duns Scotus in judging that *vera theologia est practica ... speculative igitus theologia* belonging with the devil in hell.[19] In the Protestant tradition Barth[20] identifies Baier (1686) as the first emphatically to call theology a science "presumably with reference to Aquinas", whereas others such as Gerhard "expressly rejected the designation."

With the rise of the modern universities, it is possible to track something of the trajectory of this question by considering the place of Christian dogmatic theology (as opposed to Religious Studies) within the academy. This is pertinent to our discussion in that Feyerabend is concerned that the proof of the epistemological pudding is in the power relations of society, in which academic respectability (a notion

[17] Pannenberg, 1976, p. 222ff.
[18] Aquinas, 1943, p. 6 & p. 11.
[19] See Pannenberg, 1976, p. 233ff.
[20] See CD I/1, p. 6ff.

of which he is suspicious) is an important notion.[21] Pannenberg[22] devotes a substantial part of his output on the taxonomy to this political issue, making theology's place in the academy dependent upon meeting prior criteria of scientific study, beginning with Paul de Lagarde's 1873 proposal, supported by Overbeck and taken up by Troeltsch, that theology chairs should be abolished and replaced with chairs in science of religion. Originally Overbeck had demanded that theology must renounce the presuppositions of positive revelation, but it was soon pointed out that all sciences had presuppositions. Troeltsch later countered, pre-empting Spranger's findings, that the question was not so much whether there were presuppositions, but to what extent they were open, in theory and practice, to examination. Subsequent examination of this question, would find that no clear cut distinctions between natural, historical and theological science could be made with regard to presuppositions. It is interesting that perhaps due to Harnack's opposition to these ideas, the chairs so created between 1910–2 soon migrated to the philosophy department.[23]

There are many today in English-speaking academic circles, who believe that the continuing existence of mainly Christian faculties of Theology (or Divinity) in older universities are an anachronistic inheritance of history,[24] and newer universities, if they have 'religious' faculties of one sort or another, have eschewed this tradition.

For dialectical theologians such as Barth and Diem, the requirement that scientific theology should be transmogrified into a 'science of religion' involved an intolerable distortion of focus away from the study of God, to a study of various patterns of human behaviour. Theology was a necessarily ecclesial and fideistic affair, and its inclusion in the university was a convenience and a privilege, but not one which should tempt theology to forego its proper object of study in order to secure such inclusion on more socially catholic grounds. Both theology and religious studies were *a posteriori* activities which dealt with phenomena, the facticity of which was taken as a given: the event of the revelation

[21] The thesis that a distinguishing feature of post-modernism is its belief that claims to truth are disguised attempt to justify vested interests and power is found in McGrath, 1998, p. 10.

[22] Discussion in 1976, p. 250ff.

[23] Pannenberg, 1976, p. 358.

[24] Pannenberg: "Theology is often reproached … for becoming an ideology shoring up the churches or a prevailing social system", 1976, p. 255.

of God in Jesus Christ, and human religious practices, respectively. That the object of study of religious studies was more acceptable, being universally accepted as existing and worthy of systematic study, by religious believers and sceptics alike, should not blind us to the fact that a fundamentally different object of study from theology is here being proposed, complete with a different set of assumptions and methods. Here theology is unashamed to admit with Nicholas of Amiens that it is a particular form of Aristotelian science, resting as it does on articles of faith, and therefore must resist any pressure from within that Aristotelian taxonomy which implies that the 'proper' sciences are really only those whose first principles are of a particular type (often within modernity assumed to be the empirical inductive 'type') – what makes theology scientific is, as in Aristotle's *Posterior Analytics*, not so much what type of first principles you deal in, but how you deal with them. As Torrance understands more fully than Barth, the only 'justification' for theology which does not trespass on the self-sufficient majesty of God's self-disclosure, is a bold attempt to show how theological components form part of a 'unity of truth' which incorporates all sciences: a unity which will however play out (due to a diversity of objects) as variety in terms of methods. The less theology attempts to show how it coheres with other sciences, the more its special pleading is likely to result in it being forced into an ecclesial and increasingly esoteric ghetto, where small groups of people fiddle away arranging first principles into patterns, about which no one else cares or indeed shares.

At New College, Edinburgh, which had merged with the theology faculty in 1935, Torrance found himself in the midst of these disputes, when under the Principalship of John McIntyre (1968–74), there was a decisive shift way from Christian Divinity to Religious Studies, the faculty first offering a degree in the latter in 1971. Torrance was firmly against such changes.[25]

Today objections to the retention of divinity and theology faculties are no doubt founded on the suspicion that it is a somewhat anachronistic and grubby attempt to maintain Christian exclusivism, inappropriate in an increasingly multi-cultural society. However that is unfair to those such as Barth and Torrance (and others) who argued for retention. Rather their fear is that the transition would mark a transition

[25] See McGrath, 1999, p. 104.

from the study of God to the study of religion as a human phenomenon, the latter becoming subsumed within schools of sociology, psychology, or anthropology, accompanied by an imposition of their methodologies. Those who wish to maintain the independent integrity of theology faculties are generally quite content that they should contain within them schools of Islamic, Jewish and other theologies.[26] Indeed the intimate connection between the 'science of religion' school and the study of missiology in the nineteenth century often reinforced rather than challenged a Euro-centric, and indeed often apologetically Christian, outlook.[27]

It is, therefore, inaccurate to characterize such differences merely as a battle between exclusivist and liberal interpretations of religious belief. Rather it is more accurate to characterize the differences between what Stanley Tambiah rather simplistically but effectively labels 'lumpers' and 'splitters'.[28] Griffith-Dickson[29] addresses this divide, offering the *via media* of a 'Lōkāhi' approach – taking its cue from the Hawaiian word for harmony or balance. Barth and Torrance are 'splitters' (though not exclusivists), and so is 'Feyerabend', but the exoticness of his faculties can only be construed as an attempt to dissemble the whole notion of academia.

Barth and natural science

Given that Torrance is perhaps remembered chiefly as the English interpreter of Barth, the details of his attachment to the Barthian cause are too well known to need detailing.[30] However like any interpreter, he was not simply a transmitting, but also a critical friend, who developed and sometimes rebutted Barth's ideas. On the critical side, Torrance was critical of the lingering subordinationism in Barth's Trinitarian doctrine, as well as some inconsistencies in volumes III and IV of the *Church Dogmatics*.[31] Barth is well known for his rather extreme and somewhat intemperate rejection of natural theology,[32] and most,

---

[26] For Torrance's views on the nature of the University see 1985a, p. 49ff.
[27] Pannenberg, 1976, p. 362.
[28] Tambiah, 1990.
[29] Unpublished paper: *Teaching Philosophy of religion 'multiculturally': a 'Lōkāhi' approach?*.
[30] See McGrath 1999, p. 113ff; Heron in Webster (ed.) 2000, p. 29.
[31] Torrance, 1990, p. 132.
[32] Barth, 1938, p. 6; Brunner & Barth, 1946.

including Torrance, thought that he had gone too far. In particular Torrance wrote his *Calvin's doctrine of man*[33] as a belated attempt to mediate in the famous dispute between Barth and Brunner. Torrance respects Barth's reservations, but ultimately reconstructs an analogical natural theology of a sort, which provides a 'point of contact' between the objects of theology and natural science. This divergence should not be made too much of, as Torrance regards natural theology as possible only within the orbit of revealed theology, and rejects any notion of an independent natural knowledge of God, therefore interpreting Barth's antipathy methodologically.[34] Context is important here, as Weightman is surely correct[35] in saying, that what in fact Barth is so strongly reacting to is political theology, rather than natural theology more generally. That political context Barth saw as being at a point of crisis, generally absent from Torrance's life in Edinburgh.

Finally, and for the purposes of this book most importantly, Torrance thought that Barth had failed to engage adequately with the natural sciences.[36]

Barth's reluctance to engage with natural science is then a good starting point to consider the wisdom of Torrance's more sustained involvement – was Torrance simply attempting to fill in the gaps, or was he in some way treading on ground that Barth had marked out as inherently dangerous?

In Barth's voluminous output there is scarcely any explicit engagement with natural science, there being, for instance, but a single entry in the preacher's index of the *Church Dogmatics* for natural science (CD III/2, p. 79ff – and then a 'smaller print' section). On one hand we might put this down to temperamental factors and capabilities, given that even the vast but incomplete *Dogmatics* could not cover everything. However McGrath[37] cites Barth's declining the invitation to join the 'Göttingen Conversations' in 1931, which included natural

33 Torrance, 1990, p. 127.

34 Torrance, 1969, p. 103.

35 Weightman, 1994, p. 156.

36 "However, if we are to take seriously, as Barth claims to do, the unitary interaction of God with our world in creation and incarnation, there must surely be a closer conception between the conceptualities of theological science and those of natural science, than we can find in Barth's thought. He has, I believe, pointed out to us the right direction in which to travel, but did not himself actually travel along it very far, or think it out adequately enough." Torrance, 1990, p. 152.

37 1999, p. 197.

scientists of the calibre of Howe, Heim, Weizsäcker, Heisenberg and Jordan, as an indication of a rather more active disinclination, even if the refusal was grounded in objections to Kantian preconceptions in quantum theory, rather than lack of interest in physical theory *per se*. Surely this objection cannot be taken too seriously, after all what are conversations for? It is clear even from the single entry in the *Church Dogmatics*, that a more significant reason for Barth is the fear that any engagement with natural science would mean that the it would necessarily ensnare theology "within the framework of the biologico-psychological problem which they accepted from their opponents as though it were a matter of course,"[38] and demand that it justify itself on natural science's own terms – the very thing that Barth at the outset of the *Dogmatics* refuses to give.

However if Barth shows no inclination to engage with Einstein's theory of relativity,[39] nor indeed with quantum theory – the two theories which define the outlook of post-classical physics – it is more likely to be a matter of simple chronology. Relativity theory did not receive universally accepted empirical backing until the 1960s. Quantum theory, and in particular its interpretation as a physical theory, has fared no better. Rae (2004) for instance, points out that fundamental questions of interpretation are still very much open. In Barth's time such developments were simply too new to work out in detail what their ramifications for other disciplines were. Torrance, working and writing a generation after Barth, was in a better position to assimilate and respond to the philosophical implications of these developments.

The crucial event for understanding Barth's relationship with natural science is in a clash between Barth and his 'dear friend'[40] Heinrich Scholz, which took place at the 1931 seminar on epistemology in Anselm.[41] Scholz had complained to Barth that:

> the theologian has no proof to offer himself or anyone else that he is not being merely fanciful but hears and accepts God's word. He can only feel certain. He can give neither himself nor anyone else the comfort of

[38] Barth, CD III/2, p. 85.
[39] McGrath, 2001, vol. 1, p. 176.
[40] Barth, 1930, p. 11.
[41] For discussion see Barth, CD I/1 p. 7ff; Pannenberg 1976, 270ff; Schwöbel in Webster (ed.) 2000, p. 28ff; McGrath, various, but e.g. 2002, Vol. 2, p. 285 & 1999, p. 206ff.

> legitimisation, proving that he is acting under *orders*. He can only *act* under orders and so bear witness to the presence of the command.[42]

He then set out[43] six qualifying postulates for scientific status:

1. The Proposition Postulate – the propositions of the science should obey the law of non-contradiction
2. The Coherence Postulate – "unity in its objective sphere"
3. The Controllability Postulate – that all postulates should be capable of being 'tested'
4. The Congruity Postulate – "regard to be had to what is physically and biologically impossible"
5. Independence Postulate – freedom from prejudices
6. "All propositions to be capable of being broken up into axioms and theorems and susceptible to proof on this basis"

Scholz himself knew that postulates 4 & 5 were open to disagreement even within natural science,[44] and that the claim of number 6 functioned as an overarching ('supreme') one. Barth would certainly have rejected any standard interpretation of postulate four as a 'prejudice' of exactly the sort which postulate five was designed to guard. Postulate five suffers from the defect, as Feyerabend would no doubt point out, that one person's prejudice is another's reasonable working assumption.

The main lines of disputation were therefore concentrated on the first three. Barth declares that with regard to these postulates "theology can but flatly declare them to be unacceptable to it"[45]: bridling even at the first non-contradiction postulate to which theology could only yield "upon a very limited interpretation."[46] Scholz later responded to Barth, but Barth did not reciprocate, and the dialogue ceased.

Postulate three is crucial. It looks suspiciously like the verification principle of the Vienna circle,[47] and as McGrath notes, though Scholz

[42] Quoted in Pannenberg, 1976, p. 272.

[43] In *Wie ist eine evangelische Theologie als Wissenschaft möglich?* 1931.

[44] For instance Pannenberg, 1968, p. 98: "Therefore, natural science expresses the general validity of the laws of nature but must at the same time declare its own inability to make definitive judgements about the possibility or impossibility of a singular event … the judgement about whether an event, however unfamiliar, has happened or not is in the final analysis a matter for the historian and cannot be prejudged by the knowledge of natural science."

[45] Barth, CD I/1, p. 8.

[46] Barth, CD I/1, p. 8.

[47] Though as a criteria of demarcation rather than meaning.

denied that testing and verification amounted to the same thing, he never produced any examples to the contrary.[48]

Barth's own rather briefer definition of a science is as follows:

> I propose that by science we understand an attempt at comprehension and exposition, at investigation and instruction, which is related to a definite object and sphere of activity.[49]

Other theologians have tried explicitly to meet Scholz' criteria, most notably amongst them Pannenberg.[50] Pannenberg concurs that theology can only properly be understood as a science of God, but is less willing to allow its on-going investigations into that object determine method and thereby generate distinctive theological methods.

---

[48] McGrath, 1999, p. 207.

[49] Barth, 1949, p. 9.

[50] Pannenberg, 1976, p. 270; Holwerda in Plantinga & Wolterstorff (eds.), 1983, p. 281.

CHAPTER TWO

INTRODUCING THE DIALOGUE PARTNERS: TORRANCE AND FEYERABEND

Thomas Forsyth Torrance

One of Feyerabend's most forceful challenges to natural science's own narrative, is that personal factors are just as decisive as impersonal ones in the development of a body of theory. Therefore in our account of the development of their intellectual ideas, we also admix pertinent personal details. Thomas Forsyth Torrance (1913–2007), although born in China where his parents were missionaries (that personal fact was to have important implications for his epistemology[1]), was by upbringing a Scottish Calvinist. After gaining undergraduate degrees in both classics and philosophy, and divinity, from Edinburgh University, he went to study at Basel under Karl Barth.[2] Although world events meant that Torrance was only to be in Basel for a year (1937–38), it was from that institution that he gained his doctorate in 1946 on the 'Doctrine of Grace in the Apostolic Fathers.'[3] More important was the influence that Barth was to have on him personally and intellectually, particularly through the translation (with G. W. Bromiley) of the *Church Dogmatics*. From 1950 until his retirement in 1979 he was a Professor at New College Edinburgh.[4] The focus of this book will be his work on the relationship between theology and the natural sciences. It is here that Torrance is at his most distant from Barth – perhaps more distant than Torrance acknowledged or realised.[5] His work in this field was extensive and he was awarded the Templeton Prize for it in 1978. He died on 2nd December 2007, after an astonishingly long

[1] See Torrance in Colyer, 2001.

[2] McGrath, 1999, p. 42.

[3] Torrance, 1948.

[4] Pre-empting our conclusion that Torrance does not take sufficient note of political factors in theological controversies it is worth noting in passing that initially he was appointed to the Chair of Church History for political reasons, transferring after two years to twenty seven years in the Chair of Dogmatic Theology (McGrath, 1999, p. 87).

[5] McGrath, 1999, p. 196.

and productive retirement, just as I was completing the thesis on which this book is based.

Just before his return to Scotland at the outbreak of the Second World War, Torrance was offered a post at Princeton University in the Theology Faculty by Theodore Green. The circumstances of the offer are illuminating. The university had specified the requirement that theology should be taught, not on a confessional or church basis, but 'in a dispassionate way.'[6] Torrance was precisely the former: a confessional church theologian after the manner of Barth, and in his interview headed by Green, all knew where he stood. He responded to the panel that he would teach 'theology as a science,' adding "we think not as we choose to think, but as we are compelled to think in accordance with the nature of the object, and thus in manners which are governed by the objective ground on which the science rests."[7] He concluded by pointing out that with regard to the University's demand that there should be no proselytising, he could not offer an assurance that no one would be converted by his lecturing on this basis. He was sure he had blown the interview, but the next day was offered the post, though circumstances prevented him from taking it up.

The implications of the issues raised in that interview would gestate with Torrance for twenty years, until his Hewett lectures of 1959: *The nature of theology and the scientific method.* Refined and expanded over a further decade, these were then published in his landmark book *Theological Science (1969).* The term itself derives, according to Torrance, from a series of lectures by Robert Flint of Edinburgh (a founder of the Royal Society), and is also used by Thomas Erskine.[8] Two years later there followed a sequel *God and Rationality (1971),* also *Space, Time and Incarnation, (1969a), Space, Time and Resurrection (1976), The Ground and Grammar of Theology (1980), Divine and Contingent Order (1981),* as well as many works in related fields, contributions to other volumes, and the publication of numerous articles. This is the background and material composition of the Torrance's search for a truly 'scientific theology' based around the notion of objectivity which is expounded in chapter four. It is my contention that in the notion of objectivity as *faithfulness to the nature of the object,*

[6] McGrath, 1999, p. 57–8.
[7] McGrath, 1999, p. 57–8.
[8] Torrance, 1996a, p. 237 & p. 263, as far as I can see not noted by commentators.

countering any tendency towards a prior *objectivism*,[9] we discover the intellectual heart of Torrance's project.[10]

Paul Feyerabend

Given Feyerabend's role as challenge-test dialogue partner here follows a marginally more extensive account of his 'back-story'. However again, I point the reader to chapter five as containing the core progression of ideas contained within Feyerabend's epistemological anarchy, from the theory-ladenness of observation to voluntarism, which I submit springs from a reaction to inherited foundationalist ideas within the philosophy of science.

Paul Feyerabend's (1924–1994) life is colourfully and wittily narrated in his autobiography *Killing Time*. He was born in Vienna, and after being seriously wounded in the war, studied in turn music, history, sociology, physics and philosophy. Integral to his epistemological anarchy is the diversity of his interests and examples, reflected in this variegated student diet. In 1948 he met Karl Popper at the Alpbach Summer School. Both the school itself, which he was involved in for many years, and Popper, were to be important influences in his life – of the latter, he was in turns a pupil and later a vociferous opponent. After receiving, in 1951, his doctorate from the University of Vienna on 'basic sentences', he arrived in England, hoping to study under Wittgenstein. However Wittgenstein had recently died, and so he went to be Popper's assistant at the LSE. In 1959 he immigrated to the United States, being made a Professor at Berkeley in 1962. During this period he turned away from the falsificationism of Popper, towards the more historical approaches of Kuhn and Lakatos, ending up with far more radical conclusions than either of them. He had an enduring if combative friendship with Lakatos, and *Against Method* is a product of that friendship.[11]

As Feyerabend developed his infamous later anarchic philosophy, that anarchic caste of mind began to make itself manifest in his professional and personal life. He was notoriously lax towards his academic

[9] Which by priorly conceiving an object as a material object, hardens into materialism – Putnam, 1987, p. 13.

[10] Colyer in Colyer 2001 & Colyer 2001a, p. 322, for support for this being the core of 'theological science'.

[11] Feyerabend, KT, p. 130.

commitments, eventually trying the tempers of his colleagues[12] and university authorities.[13] His first and best known book by far is *Against Method* (1975, revised 1988 and 1993). Although his first full book, it marks a complete and radical break with his earlier Popperian approach,[14] and was to be followed by other books: *Science in a free society (1978), Farewell to reason (1987), Three Dialogues on Knowledge (1991);* two volumes of collected papers: *Realism, rationalism and the scientific method (1981), Problems of empiricism (1981),* the already mentioned autobiography *Killing Time (1995);* published posthumously were *Conquest of Abundance (1999), Knowledge, science and relativism, (1999)* and his correspondence with Imre Lakatos *For and against method (1999).*

Until *Against Method,* Feyerabend was known as a regular, if increasingly radical, philosopher of science. Subsequently he forewent the usual restrictions on a philosopher of science, and like Popper, increasingly focused his attention on the broader question of the place and influence of natural science within our political and social culture. Consequently he was taken less seriously by philosophers of science, for whom his work was perceived as lacking the necessary focus and rigour, but therefore gained a wider, rather radical left wing audience under what we might today loosely call the post-modern umbrella.[15] He describes his own 'programme' to Lakatos as "abandon scholarship, construct a new myth, and present it with the help of a movie."[16]

Unlike Torrance, Barth and Popper, there is therefore only a self-limited sense of cohesion in Feyerabend's work. By design there is no rolled-out 'programme', and even his *volte face* is no smooth transition.[17] There were elements of Feyerabend's thought which developed early on, gestating in non-anarchic form, which would, sometimes decades later, form vital parts of his epistemological anarchism. For instance, the first published version of his incommensurability thesis he dates to 1958,[18] and discussions of it from 1952.

---

[12] See Hoyningen-Huene in Preston, Munevar and Lamb (eds.), 2000, p. 7.

[13] He held several appointments, subsequent to California, sometimes concurrently: University College London (66–69), Free University of Berlin (68–70), Yale University (69–70), University of Auckland (72–74), University of Brighton (74–75), University of Kassel, Zurich (1970–90).

[14] Preston, 1997, p. 11.

[15] Preston, 1997, p. 7.

[16] Feyerabend, in Feyerabend and Lakatos, 1999, p. 262.

[17] Munevar in Preston, Munévar, & Lamb (eds.), 2000 p. 58.

[18] See AM, p. 211.

Crucially therefore, Feyerabend's intellectual anarchism, which affects his philosophy from the late 1960s, is embodied not only in what he says, but in how he says it. He introduces subjects and principles, and then drops them; he changes his mind, and then changes it back again; he deliberately takes both sides of an argument; he is even blatantly contradictory, after all for the anarchist, "there is nothing wrong with inconsistencies."[19]

Feyerabend so positively delights in self-rebuttal, inconsistency, irony, provocation, posing, devil-advocating, u-turns, and so loathes the (so-serious) programmatising output[20] of chiefly Karl Popper, that not only are we likely to get caught out if we take remarks too seriously out of context, but also are likely to miss the fact that the method is in the madness – that it is irony, inconsistency, provocation, which are in themselves what Feyerabend as a philosopher is about. An anarchic 'programme' is a contradiction in terms. He is ironic and humorous in his writing, and delights at the pomposity of those outraged commentators who take his pronouncements seriously.[21] Though conversely he could also be rather defensive and was so genuinely hurt by the hostile reaction to *Against Method,*[22] that he responded at some length and with ferocity to all those who had voiced criticism.[23]

His 'movie proposal' is therefore to be taken seriously, if not literally, as a metaphor for a dialogical method that is dynamic, witty, visual, narrative, contrapuntal, and contains its fair share of false starts and dead-ends. The other metaphor that Feyerabend himself employs is that of 'collage'.[24] The elements of the collage, Feyerabend says, are to be found in the preceding twenty years of his writing. In a collage these elements are drawn together, not in a cohesive manner, but in the disconnected manner of the archaic style, a style of paratactic aggregates from which Feyerabend was fond of drawing inspiration.[25] Having championed the archaic style, Feyerabend consequently considered himself under no obligation to place 'idea-objects' in perspective in his works.

---

[19] Feyerabend, SFS, p. 191.

[20] Preston in Preston, Munévar, Lamb (eds.), 2000, p. 10. Also Feyerabend, TDK, p. 123 – Feyerabend praises Lessing in this regard.

[21] Feyerabend, AM, p. vii.

[22] For instance, Agassi, 1976; Floyd 1991; Harré, 1977.

[23] Preston, 1997, p. 170.

[24] Of AM, Feyerabend, KT, p. 139.

[25] Feyerabend, AM, p. 170ff; FTR, p. 97ff.

Any commentary therefore on Feyerabend must make clear which period of his life it refers to, and even then beware. We shall be considering the later Feyerabend, as he develops the ideas in *Against Method*, both in the years before its writing and up until his death.

Published posthumously, *For and Against Method*, (the correspondence between Feyerabend and Lakatos, whose sudden death in the midst of the project angered and shocked Feyerabend) is a partial reconstruction of the original intent of *Against Method*, a project at its outset conceived of as a dialogue between Feyerabend and Lakatos. The actual *Against Method* should, insists Feyerabend, also be understood in that light – as one side of a conversation between friends. Its iconoclastic controversialism was its main selling point and also the reason it attracted more or less universal hostility from fellow philosophers of science – as Feyerabend himself was to say: it was "a stink bomb."[26]

The backdrop to Feyerabend's latter work, personally and culturally, is the logical positivism and Popperianism of his earlier work, a backdrop shared with Torrance, and the work of both can be viewed as a reaction to the excesses of the former. The logical positivism of the Vienna Circle, associated with names such as Schlick and Carnap, and in England Ayer, was by far the most influential movement in the philosophy of science in the first half of the twentieth century.[27] Its rise and fall is too well known to need much comment. Its main problems were threefold.

Firstly, the arbitrariness of the verification principle, however it was drawn up, was always liable to be self-refuting.[28]

Secondly, the sharp demarcation between observational and theoretical statements which logical positivism demanded, could not be maintained. This is a key casting off point for Feyerabend's epistemology.

[26] Feyerabend, KT, p. 147.

[27] Ian Hacking describes the family resemblances which demark positivism: "positivists are empiricists who typically emphasize verification, falsification and observation; they feel uneasy about causes, explanations, theoretical entities and metaphysics. Their opponents, scientific realists, want to think of theoretical entities as real, of causal relations as necessary, and of metaphysics as acceptable proto-scientific explanation." (Preston, 1997, p. 31).

[28] Popper, 1992, p. 36. Torrance quotes similar passages from this book in his dismissal of positivism in Torrance, 1969, p. 168; see also Popper, 1969, p. 253.

Thirdly, though some practising scientists, such as Mach,[29] were strongly positivist in their views, there was considerable doubt as to whether in fact scientists did or should work as logical positivists wished them to.[30]

Theologically the movement was important, because however construed, the verification principle invariably consigned statements about God to the metaphysical rubbish heap, A. J. Ayer infamously dismissing discussion of the knowledge of God in two and a half pages.[31]

However the demise of logical positivism, as spectral as its rise, has in turn, offered theology at least an opportunity for philosophical rehabilitation.[32]

Popper sought to supersede logical positivism by the replacement of the verification criterion with his falsifiability criterion, though crucially this criterion was to function as an epistemically-neutral demarcator of science, rather than of meaning.[33] For theology this entails that its metaphysical propositions are restored to meaningfulness, but being generally regarded as falling on the wrong side of this new line of demarcation, falsificationism continues to deny the possibility of a theological *science*.

However Popper's falsification criterion itself has several inconsistencies. Firstly, the Duhem-Quine thesis shows the impossibility of isolating one particular theory for testing; secondly, Lakatos' study of the persistence of falsifiers as anomalies[34] showed that anomalies were even present as the theory was proposed;[35] thirdly, in continuity with logical positivism it maintained an ultimately untenable distinction between theory and observation, much criticised by Feyerabend.

It is ironic that the epistemological anarchist Feyerabend, after his anarchic revolt against Popperianism, blames the academic and personal breakdown in relations between Popper and himself on *Popper* having changed. The Popper of Alpbach – "iconoclastic, humorous, journalistic" – who started his lectures by stating that though a

[29] Though Feyerabend disagrees with this interpretation of Mach – see p. 127.

[30] Ritchie, 1948, p. 22 & Feyerabend, FTR, p. 190.

[31] Murphy, 1990, p. 53.

[32] Popper, 1992, p. 51, cited Torrance, 1969, p. 167f; Plantinga in Plantinga & Wolterstorff (eds.), 1983, p. 21; Polanyi 1998, p. 9.

[33] Popper, 1992, p. 311.

[34] Lakatos, 1978, p. 3, also in Feyerabend & Lakatos, 1999, p. 81; Polanyi & Prosch, 1977, p. 146; Torrance also notes this: 1985, p. 49.

[35] Lakatos in Feyerabend & Lakatos, 1999, p. 83, with regard to Newton's *Principia*.

Professor of Scientific Method, there was no such method but just "some simple rules of thumb"[36] – had become, according to Feyerabend, a far more serious and establishment character.[37]

Feyerabend's reading, in draft form, of Kuhn's *The structure of scientific revolutions*, was to mark another important turning point in his thought. Kuhn's historical approach, was to occasion a similar shift in Feyerabend from a prescriptive and normative approach to the philosophy of science, to a more historical and descriptive one.[38] Philosophy of science could no longer aim to describe 'good science' from a set of abstract principles, but instead must rather more humbly, describe patterns within past developments and make inductive recommendations for the future.

Imre Lakatos, Feyerabend's dearest male friend,[39] and the foil of *Against Method* labels Feyerabend "our brilliant leading cultural relativist"[40]: a label which will be important in our assessment of Feyerabend's epistemology. Lakatos' basic unit of scientific activity is the 'research programme' which contains a 'theoretical core', and a 'protective belt' of auxiliary theories, the latter open to empirical investigation and refutation. A programme in which the protective belt is growing, and being absorbed into the core, is labelled 'progressive' (a good thing according to Lakatos, but not Feyerabend) and those where empirical refutations are threatening the core, 'degenerating'.[41]

[36] Feyerabend, KT, p. 88.
[37] Feyerabend, TDK, p. 50.
[38] Feyerabend, TDK, p. 156.
[39] Feyerabend, KT, p. 130.
[40] Lakatos, 1978, p. 178.
[41] "A research programme is said to be *progressing* as long as its theoretical growth anticipates its empirical growth, that is, as long as it keeps predicting novel facts with some success ('*progressive problemshift*'); it is *stagnating* if its theoretical growth lags behind its empirical growth, that is, as long as it gives only *post hoc* explanations either of chance discoveries or of facts anticipated by, and discovered in, a rival programme ('*degenerating problemshift*'). If a research programme progressively explains more than a rival, it 'supersedes' it, and the rival can be eliminated (or, if you wish, 'shelved')." (Lakatos, 1978, p. 112)

Feyerabend's criticism of Lakatos hinges on this final ambiguity, and also in an initial ambiguity of what constitutes a research programme (not defined by Lakatos, in the same way that the precise meaning of the term paradigm is so problematic for Kuhn (Marty in Küng and Tracy, 1989, p. 182,) suffice to say Lakatos' term is a much larger scale unit than the everyday research programmes which natural scientists are used to referring to when applying for grants). Feyerabend tends to conceive of research programmes as very general theories (indeed he increasingly calls them 'traditions') of very large scope: 'atomism' is the paradigmatic example (see FTR, Ch 5–7). He then asks whether the difference between 'eliminated' and 'shelved' is not a vital

Feyerabend, like Popper,[42] traces the roots of the Western 'modern' scientific tradition back to the pre-Socratics' sceptical overthrow of the mythical Homeric hegemony (in contrast to Torrance who, as we shall see, emphasizes the discontinuities between modern physics on the one hand, and classical Newtonian physics and its Greek inheritance on the other, choosing, somewhat eccentrically, Calvin as the originator of "modern scientific thinking" – see p. 231). Feyerabend's particular favourite is Xenophanes,[43] though his Xenophanes is very different from the one whom Popper championed.[44]

Feyerabend, with typical perversity, describes Xenophanes not as a champion of Popperian 'rationalism', but as a "conceited big mouth" who used 'charm' with 'one-liners' to conceal a lack of substance in his arguments (Aristotle called him "somewhat uncouth").[45] His victory and the victory of the pre-Socratics was not a victory of reason over myth – for Feyerabend reason itself is a device used to support myth – but a victory for individual freedom over hegemony. Here we find the key to understanding Feyerabend's unusual interpretation of scientific history. Myth is defined, notwithstanding any technical usage within hermeneutics and elsewhere, simply in terms of political and cultural hegemony – a belief system or scientific paradigm is judged

---

one, if natural science as a whole is to be classed as 'progressive'. The definitive rejection of antiquated research programmes would vouchsafe for natural science's progression. However if research programmes are only 'shelved', albeit indefinitely, then the question remains open as to when and if it is appropriate to take them off the shelf and dust them down. For Feyerabend it is thus possible to understand 'atomism' as referring to a part of Epicurean philosophy which is dusted down by Gassendi, and subsequently absorbed into modern atomic theory, despite approaching two millennia of obsolescence. The opposing view is that they are two separate programmes, with merely a common name and passing resemblance. If there is constantly a chance at least, that any long 'rejected' programme might re-emerge, then this undermines natural science's claim to be progressive. The possibility that, for instance, the theories of Leucippus and Democritus, and Rutherford can be seen as part of the same 'research programme' is a sign of the elastic definition of this term.

Murphy and Laudan (1977, p. 83) also employ Lakatos' schema, though the latter makes the important modification that the theoretical core need contain no elements which are constant over time. Therefore Laudan makes clear that shelving is the strongest form of rejection possible, together with the stipulation of the conditions upon which it might be rehabilitated.

[42] 1969, p. 68ff.

[43] Feyerabend, FTR, p. 90ff.

[44] Feyerabend, FTR, p. 94 – contrast with Popper, 1969, p. 226, Xenophanes is guided by a "theory of objective and absolute truth".

[45] Feyerabend, FTR, p. 91.

mythical solely because, politically and culturally, its status has become unassailable. For Feyerabend, the hegemony of modern science, whilst it might justify its claims with reference to such qualities as rigorous method, solid empirical evidence, and sophisticated mathematics, it is in fact just as 'mythical' as the Homeric myths – purely on the grounds that it enjoys a similarly privileged status in society. If Kuhn ultimately reduces truth to a show of hands, then his anarchic out-rider Feyerabend (who has little time for the concept of 'truth' as such) gives priority to those minority theories and traditions (the epistemological underdogs of their historical contexts). The desired scientific characteristics of the champions of those underdogs are therefore tenacity in the face of power and the pursuit of freedom in the face of totalitarianism.

Furthermore Kuhn's descriptive account of science had identified the two stages of normal and revolutionary science. Prescriptively most, including Kuhn himself, had argued, or more often simply assumed, that a scientific paradigm ought to strive to attain 'normal' status.[46] Contrastingly Feyerabend's prescription is constant revolution.[47] Lauden likewise points out that Kuhn's and Lakatos' rationalism suffers from the defect of not specifying why the normal *should* be preferred to the revolutionary.[48] The true *scientist* is, for Feyerabend, the one who uses *whatever is necessary* tenaciously to promote an unlikely theory – the prophet, the revolutionary and indeed the propagandist. *Reason* according to Feyerabend, is to be distrusted because it is a weapon more commonly procured by the establishment than the freedom fighter.

If it were to be therefore possible to speak coherently about a tradition called 'Western natural science', then, according to Feyerabend, within that tradition there will be points at which its normal hegemony[49] will render it 'mythical', and at those times of 'revolution' it will regain its true identity as the subverter of the prevailing myth.[50] Feyerabend is perfectly happy therefore to label natural science as

[46] See for instance Foucault's identification of normal science as 'non-dubious' science – Dreyfus & Rabinow, 1982, p. 10.

[47] Preston in Preston, Munevar & Lamb (eds.), 2000, p. 91.

[48] Lauden, 1977, p. 151.

[49] Whilst Timothy Gorringe's book on Karl Barth is concerned with the social and political elements of Barth's theology, extraneous to the concerns of this thesis, we simply note its title: *Karl Barth: Against Hegemony*, OUP, Oxford, 1999.

[50] Feyerabend, AM, p. 295.

"our little fairytale,"[51] for "there is no distinction at all between science and myth except perhaps the date."[52]

Here therefore we have the conflation of the epistemic, ethical and political requirements which determines Feyerabend's conception of the 'scientific'.

Politically: the belief that individual freedom ought to remain sovereign against the centralising and totalising tendencies of the state.

Ethically: the belief that such freedom accrues rights to the individual and to minority groups, to live according to their own traditions, including when those traditions are informed by an 'unscientific' worldview.

Epistemically: the belief that the voice of dissenters and outsiders should be taken seriously and on their own terms; and that scientific theories, insofar as they are incarnated in educational procedures and research institutions which demand public monies and academic privileges, are implicitly political and ethical, so that Feyerabend can talk of scientific theories, being in and of themselves 'oppressive'.

This conflation has epistemic consequences: "This liberal practice [democratic freedom with regard to how public money funds scientific research, and what can be taught in public schools], I repeat, is not just a *fact* of the history of science. It is both reasonable and *absolutely necessary* for the growth of knowledge."[53]

Politically Feyerabend stands in the liberal tradition, which he traces to Mill's 'immortal thesis'[54] *On Liberty*. Whilst it is clearly beyond this book to assess the standing of this liberal tradition, the connections which Feyerabend makes between the ethical and political on one hand and the epistemic on the other are unusual. We are not here to argue the case for or against liberalism *per se,* but to note Feyerabend's conflation of social stances and motives, and epistemological outlook.

Most significantly these linkages are for Feyerabend *inevitable and necessary*. It is a commonplace to say that political regimes have often sought to use, limit and manipulate the autonomy and prestige of the scientific community, and that the easiest leverage it has had, has been through the allocation of public funds. Scientific patronage has always

---

[51] Feyerabend, AM, p. 298.
[52] Feyerabend, KSR, p. 60.
[53] Feyerabend, AM, p. 14 (his italics).
[54] Feyerabend, FTR, p. 33.

been necessary and has always been accompanied by the suspicion that he who pays the piper calls the tune. Such patronage has been pursued by a variety of agencies which have had and continue to have their own economic and political agendas, and of course also their own theological agendas. The case of Galileo is not only the story of the clash of scientists and theologians but also the story of a clash between scientist and patron, and according to Feyerabend it is the latter which is decisive. Indeed one of the problems of unravelling the case study of Galileo is that theology and science were so intertwined as disciplines. Whilst such a relationship might be problematic in itself, it cannot be ignored or discounted in historical analysis.

Scientific patrons have been individuals, states and increasingly commonly private corporations. Each of these is likely to have its own agendas, and its own mechanisms of control and manipulation, some more open to scrutiny than others. The involvement of armament manufacturers, tobacco companies and pharmaceutical giants in scientific patronage are but three contemporary examples which arouse heated debate. Generally the response of philosophers of science has been to outline prescriptive strategies by which science should clean up its act: the internationalisation of research, the open declaration of sources of funding, the independent peer review of all research, the autonomy of scientific institutions, publication of raw data, and acknowledgement of methodological assumptions, have been the standard defences against political or economic interference in natural science. Feyerabend believes that such defences can only ever amount to window dressing, and that the connections between money, power and results are inevitable.[55] *If we can't get the politics out of science, Feyerabend is saying, then we ought to concentrate our efforts on ensuring that that politics is humane* – any other course of action, however scientifically high-minded, is liable to be subverted. Therefore anarchy is the only medicine[56] strong enough to challenge the hegemony of Western science, which is an inherently colonising beast.

With some justification Feyerabend claims that transitions from revolutionary to normal phases has denied natural science its 'outlaw' status, which provided its initial impetus: "Twentieth century science has given up all philosophical pretensions and has become big

[55] Feyerabend, AM, p. 66.
[56] Feyerabend, AM, p. 17.

business. It no longer threatens society, it has become one of its most powerful supporters."[57]

* * *

As a child Feyerabend was precociously punctilious in his religious observance, often going to confession two or three times before Holy Communion.[58] However such youthful enthusiasm, gave way to an equally enthusiastic positivist rejection of religion.[59] Although Feyerabend later again changed his mind, and developed a certain 'solicitude' for religious thought, there is no evidence that the mature Feyerabend was a practising or private member of any religious tradition. His final verdict on religion was "all religions are good 'in principle' – but unfortunately this abstract Good has only rarely prevented their practitioners from behaving like bastards."[60]

References to the 'church' in his writings, which are conspicuously numerous, are nearly all found within an abusive and oft-repeated metaphor for what natural science has become.[61] The prestige, authority, and money which society presently invests in natural science is exactly that which in the past it had invested in the church: "The judgement of the scientist is received with the same reverence as the judgement of bishops and cardinals was accepted not too long ago."[62]

So just as there are those who believe that the efforts of those in nineteenth century mission fields were hopelessly compromised by their dependence upon their imperialising compatriots, even if those relations were sometimes sharply critical, so natural science today is similarly compromised today by its paymasters:

> There was no 'objective' comparison of methods and achievements. There was colonization and suppression of the views of the tribes and the nations colonized. These views were replaced, first, by the religion of brotherly love, and then by the religion of science.[63]

[57] Feyerabend in Feyerabend & Lakatos, 1999, p. 114, see also Grunefeld, 1996, p. 317.

[58] Feyerabend, KT, p. 19.

[59] Feyerabend, FTR, p. 264.

[60] Feyerabend, FTR, p. 299.

[61] E.g. Feyerabend, AM, p. 46; RRSM, p. 76; KSR, p. 182; SFS, p. 74; also a metaphor shared by Polanyi: 1998, p. 265; 1969, p. 40; Scott, 1995, p. 12.

[62] Feyerabend, KSR, p. 182. See also Feyerabend, SFS, p. 74; Feyerabend, AM, p. 46; Feyerabend, RRSM, p. 76. Gelwick concurs: "Science exercises the kind of authoritative sway that Christian religion once did. To assert that a person, or his idea, is 'unscientific' is the severest charge" (1977, p. 14).

[63] Feyerabend, SFS, p. 102.

Thus Feyerabend is often surprisingly kind about the present day church, particularly where it clashes with natural science, not through any religious conviction, but because Christianity has once again become an 'outsider' whose right to have its say must be reasserted. However the implication if we reverse the metaphor is equally clear: should Christianity make a renewed attempt to claim for itself objectivity (as *per* Torrance), it would again fall under renewed Feyerabendian censure. Religion and natural science are in the same boat, neither can fully shake off their imperialistic histories, and carry on regardless, each must be reconstructed in a more politically acceptable form, regardless of the consequences for 'objectivity'.

Feyerabend wishes to moderate such vested interests by a system of direct democratic control over both funding priorities and educational practices. "A free society is a society in which all traditions are given equal rights, equal access to education and other traditions of power."[64] Democratic citizens' councils should determine the flow of public funds within the scientific community,[65] and parents should choose what is taught to their children in schools, with options such as voodoo and legends on the curriculum. Feyerabend is not much bothered by the practicalities of his suggestions,[66] though he offers the example of trial by jury, as a similar situation in which ordinary citizens make important judgements over the arguments of experts.[67] More importantly neither is he concerned with any consequential epistemic weakness: "Participation of laymen in fundamental decisions is therefore required *even if it should lower the success rate of the decisions*."[68] However earlier we caught Feyerabend saying that such a 'liberal practice' was "*absolutely necessary* for the growth of knowledge." Feyerabend cannot hedge his bets here: either theoretical pluralism is necessary for the growth of knowledge or it should be pursued regardless of its epistemic success rate – these two grounds for justification of this pluralism are quite distinct and cannot be coalesced.

Although he contrasts these ideas with those of Polanyi, one must remember the context[69] – Polanyi's defence of science's independence

[64] Feyerabend, SFS, p. 30.
[65] Feyerabend, TDK, p. 120.
[66] Newton-Smith, 1990, p. 133.
[67] Feyerabend, SFS, p. 95.
[68] Feyerabend, SFS, p. 87, his italics.
[69] Gelwick, 1977, p. 40.

was with reference to totalitarian interference in science, Feyerabend was concerned with the place of science in a liberal democracy.

Feyerabend's ideas are also subject to the classic liberal dilemma of what to do with 'traditions', such as anti-semitism which are fundamentally illiberal. Feyerabend gladly impales himself on one of the horns: "traditions are neither good nor bad, they simply are. Objectively speaking, i.e. independently of participation in a tradition, there is not much to choose between humanitarianism and anti-Semitism."[70] Quite how it is possible or desirable for a liberal to use the 'anthropological method'[71] to 'participate in' anti-semitism is not discussed by Feyerabend. His suggestion of self-electing communities of sado-masochists[72] does not cover this situation where the victims are anything but self-selecting.

As Feyerabend's written output began to transcend the normal boundaries assigned to the philosophy of science, and encompassed areas more normally thought of as part of the social sciences (not that Feyerabend ever felt over-concerned by any such disciplinary demarcations), so the material which he analysed increased in scale – from scientific theories, to paradigms (Kuhn's favoured term) and latterly to the somewhat loosely defined notion of 'traditions'.[73] The tradition of Western natural science he understood to be particularly powerful within our own culture,[74] but ultimately one which needs to take its place alongside the traditions of other cultures.[75] Christianity, for Feyerabend, is simply portrayed as another competing tradition, and one which in the cultural-political clothing of Christendom, had largely had its day. The power of bishops and cardinals lies in the past, thus as unfashionable outsiders, he develops a certain 'solicitude for theology', but the assumptions behind such solicitude may not however be cause for "churchmen to sleep easier in their beds."[76]

Feyerabend wishes that these traditions should engage in 'exchanges'[77] with one another, and that those exchanges should be 'open' rather than 'closed' exchanges – ones in which the rules of the

[70] Feyerabend, AM, p. 225.
[71] Of Evans-Pritchard, see AM, p. 189.
[72] Feyerabend, SFS, p. 132.
[73] For instance, AM, p. 225f – a usefully flexible term.
[74] Newton-Smith, 1990, p. 1.
[75] Popper, 1969, p. 129.
[76] Watkins in Preston, Munévar, Lamb (eds.), 2000, p. 56.
[77] Feyerabend, AM, p. 269.

exchange are not determined beforehand or exclusively by either party. There therefore can be no consistent and universal characteristics by which open exchanges are identified or with which they are conducted. However, Feyerabend argues that when such exchanges have involved Western natural science on one side, they are *de facto* closed exchanges, because that tradition has by virtue of its inherent imperialism 'rigged' them, claiming that its own standards of rationality and objectivity must be adhered to – whereas in fact those standards themselves, far from being 'objective' are simply those which act in its favour.[78] Part of the narrative of this tradition includes the claim that it has identified universal characteristics of predicates such as 'rational' and 'scientific'. This however involves Feyerabend in the usual relativist regress: how can we show there are no such things as 'objective' criteria of assessment without the necessary 'meta-criteria' by which such instruments might be recognized?

The effect of this is that on the occasions when Feyerabend does comment upon the relations between natural scientific and theological traditions, he lives firmly within the 'conflict' model, and even if he exonerates theology, this is of little help to theologians like Torrance, who wish the exchange to be a little less hostile and a lot more complex. Torrance (nor I) does not necessarily wish to side with Californian fundamentalists in their 'exchanges',[79] however much Feyerabend might thus wish to offer such religious traditions an epistemic helping hand.

At the risk of here forming a premature conclusion, it appears that Torrance on the other hand, by showing the damage done by scientistic conceptions – damage to *natural science as well as* theological science – demonstrates the path to a more mutually fulfilling and less combative dialogue. Feyerabend is too concerned with showing how natural science *has* 'closed' exchanges, to offer examples of how open exchanges might be possible, or indeed where they might have occurred or are occurring. Feyerabend's accusation levelled at theologians, such as Bultmann, of, in effect, surrendering prematurely[80] to natural science's demand for closure – is a charge which other theologians, not

---

[78] Feyerabend, SFS, p. 102.

[79] " "Three cheers for the fundamentalists in California who succeeded in having a dogmatic formulation of the theory of evolution removed from the text books and an account of Genesis included." (Feyerabend KSR, p. 187).

[80] Feyerabend, KSR, p. 143: 'castrating' is Feyerabend's typically colourful term.

least Torrance and Barth would agree. Furthermore Torrance provides, in some depth, just the sort of 'open' exchange which Feyerabend demands but never provides. Indeed given that Feyerabend's thesis is undermined by such provision, he has no interest in pointing it out – rather he wants to stick with his conflict model, in order to "teach them [natural scientists] some manners."[81] The monolithic facticity of his traditions ("they simply are"), by its inherent and dogmatic relativism, leaves dialogue just as hamstrung as a retreat into fundamentalism, naturalistic or theological. Whilst the democratic consensus may well provide an ethical mechanism for making decisions about funding research, unless we adopt a Kuhnian 'show of hands' approach to truth (which Feyerabend repudiates directly with his theoretical pluralism), we are left with little ability to assess traditions epistemologically. This is because 'truth' has, for Feyerabend, become a purely pragmatic and suspect quality, ripe for deconstruction.

Our rather premature conclusion is therefore that 'traditions' are not the solitary homogeneous monoliths which Feyerabend assumes. Certainly they can be, but they can also develop, mutate, syncretise, split, shift and absorb in manifoldly interesting ways. Put simply, 'open exchanges' can and do occur, even ones including natural science, and yet Feyerabend, for his own reasons, chooses to ignore them – indeed they are the mechanisms by which ever-evolving 'traditions' emerge.

Underlying Feyerabend's rather clumsy approach is his championing of 'the anthropological method' of Evans-Pritchard[82] as a means by which we might assess 'scientific traditions',[83] predicated upon anthropological field work in which a people or tribe can be isolated and defined.[84] Ironically for Feyerabend, this method is a universal method elevated above others.[85] The problem is that the 'method' has its shortcomings and doesn't work well everywhere – in particular in existing cosmopolitan communities, because one of its preconditions is that the encountered tradition is suitably 'closed', thus including the conclusion which Feyerabend is after, in the premises. Feyerabend

[81] Feyerabend, FTR, p. 260.

[82] Feyerabend, FTR, p. 267f; AM, p. 189.

[83] He is not explicit how this might for instance allow a modern physicist to indwell Druid or Aristotelian cosmology.

[84] Furthermore as Nielsen points out (in Nielsen & Phillips, 2005, p. 150) mutual comprehension is just as likely to occur within unhomogeneous societies as it is between them, even when everyone – notionally at least – speaks the same language.

[85] Preston, 1997, p. 188.

attributes science's 'success' to its ruthless imperialism, but another commentator might see things the other way around: science is successful because, particularly when articulated in the international language of mathematics, it is supremely culturally flexible and exportable, so that a wide diversity of people can participate in and benefit from it, certainly by accommodating culturally inherited beliefs, but not necessarily simply abandoning them. A modern cosmopolitan and technology city can accommodate a high amount of cultural diversity. Certainly the economically powerful have often tried to enrol science in the struggle to preserve their privileges, but for that enrolment to be necessary a case must be made. However because of its wide cultural penetration it is not easy to make the case by considering science as a singular cultural tradition and employing the anthropological method.

Whilst it is one thing to adopt the language, customs, and even, imaginatively, the world view of a particular tribe, that does not entail that we transcend our position as observer, outsider, and interpreter. It is an experimental technique not a means of metaphysical transformation. Our position as observer is not transcended but underlined – it becomes a particularly involving 'thought experiment'. We cannot transcend our own background to the extent that we cease to evaluate the host traditions by the standards of our own.[86] Furthermore, Feyerabend himself believes (as we shall later see with 'natural interpretations') that many of these standards are unconscious to us, so that even as we strive to divulge ourselves of our cultural prejudgements, the most covert[87] and damaging of these remain unquestioned and unacknowledged. Is not the whole notion of the anthropologist as scientific observer of humans, a Western-constructed activity?[88] Is this not in fact a false subjectivity, which is the twin sister of Enlightenment objectivity? Is not actually the proper mode of interaction the more subtle realignment of subjectivity and objectivity which (as we shall see) Torrance advocates – the way of honest dialogue, MacIntyre's 'dialectically constituted traditions', which acknowledges both where our own tradition stands, and is open to the possibility of the other?

[86] See e.g. Nielsen in Nielsen & Phillips, 2005, p. 25. There is a parallel difficulty in hermeneutics here, and for instance in authentic performances in the arts – how can we (e.g.) become C18th Viennese listeners when attending an 'authentic performance' of Mozart.

[87] To use a term of Whorf's.

[88] Hacking, 2000, p. 211.

However if Feyerabend, for his own ends, is keen to cast all Christians as fundamentalists, it is also the case that theologians (and other members of the humanities) are liable to cast natural scientists as positivists of one hue or another. Like Feyerabend, there is no doubt that Torrance had spent considerable time studying and interacting with the natural science around him.[89] The danger is that theologians and social scientists, who have little knowledge or experience of how natural science actually operates, are cowed by its cultural success, or harness themselves too easily to the safety of a particular philosophical tradition. Foucault for instance,[90] accepts without question the Kuhnian distinction between 'normal' and 'revolutionary' science, assumes that the former is a desirable and identifiable state of scientific maturity (whereas Feyerabend would dispute both those qualities), and refers to natural science as 'non-dubious' as opposed to the 'dubious' findings of social science, without any nod towards the critique that those such as Feyerabend might make. Torrance takes a more robust and measured line with both Christian theology and natural science.

Thus both Torrance and Feyerabend agree that the 'traditions' of theological and natural science cannot avoid, and indeed should, enter into dialogue. Furthermore both agree that theology has recently been too timid in its engagement. However where they differ, is that Feyerabend, in order to bolster his own misgivings about the present dominance of natural science, cannot conceive of Christian theology meeting natural science in any mode apart from the confrontational. Whilst he appeals for open exchanges, he is too busy berating natural science for its rigging of such exchanges, to imagine or identify those places where such an exchange occurs. I believe that whilst not perfect, Torrance is at least attempting such an exchange,[91] wherein the integrity of both natural and theological science is upheld.

Feyerabend as critical friend to Torrance

There is no doubt that Feyerabend and Torrance were personally very different sorts of people, given that they were both Western academics.

---

[89] For good measure McGrath, Barbour & Polkinghorne – it is notable that few professional scientists who turn to theology embrace demythologising with Bultmannian enthusiasm.

[90] Dreyfus & Rabinow, 1982, p. 10.

[91] Torrance, 1969, p. 10 – 'open epistemology'.

There is both danger and utility in dealing in stereotypes. Feyerabend, stereotypically, was the freewheeling West coast American 1960s academic philosopher: epistemological anarchist; free thinking and iconoclastic; not wishing to found or be identified with any school of thought[92]; a subverter of tradition; personally something of a hedonist – with an eye for a beautiful woman; a lover of wine and beer, as well as of the arts, in particular opera, where he might have made a living; rootless and restless, professionally and personally.

Again, stereotypically, we might cast Torrance as the rather dour Scottish Presbyterian theologian: son of missionary parents; married to one woman; family man[93]; spent most of his life at one University, indeed most of it in one post at that institution; as a thinker saw himself as contributing a tiny fragment of the great classical tradition of Christian theology; apart from a scholarly interest in natural science, little indication of other interests outside his discipline.

Although Torrance outlived Feyerabend by some twenty four years, both were writing at about the same time, with key publications dating from the decade starting 1969. Both were writing about the place of natural science in relation to other elements of academic and popular culture. Both sought to respond to the intellectual climate of the time: whilst noting the simplistic and monochrome nature of the portrait, they were writing to a world in thrall to the rise and progressiveness of empirical science and particularly the technological marvels[94] it brought in its wake.[95]

Within the philosophy of science the high-water mark of scientism – logical positivism – was already on the wane, and the more open approach of Popper's falsificationism was also being subjected to substantial criticism. Within philosophy and culture more generally, the most important tenets of Enlightenment thought – classical foundationalism, the primacy of reason, and a belief in universal progress[96] – still held varying degrees of support, though the dissenting voices – always present – were growing more confident. Each

[92] Feyerabend, TDK, p. 123.

[93] For all his scholarship, this was his self-confessed greatest joy and achievement: Colyer, 2001a, p. 13.

[94] Feyerabend, SFS, p. 175.

[95] Their fairly precise contemporaneity is an important factor in considering them together, their landmark books appearing within six years of one another.

[96] Dilthey, quoted in Griffith-Dickson 1995, opening quotation.

therefore had different motives and objectives in their writings, but each were responding to a common cultural and academic milieu.

Torrance saw an opportunity within a more conducive intellectual climate – made so by shifting paradigms within natural science, and less scientistic approaches in the philosophy of science – to reassert the epistemic potential of a Barthian Christian theology based on divine revelation. A less positivistic conception of science could be formed around a new form of objectivity. Feyerabend saw such changes as demanding a more wholesale re-evaluation of the place of science both within the academy and society more generally. For him the demise of Enlightenment foundationalism demanded a more radical deconstruction of concepts such as rationality and objectivity. Feyerabend is generally understood to be one of the early prophets of the umbrella-movement known as post-modernism,[97] even if that term continues to be somewhat abused and ill-defined.

Torrance could therefore, as I have already said, be assessed straightforwardly with reference to his own context. The disadvantages of such an assessment is that firstly some work of that kind has already been done, secondly 'the context' would have to be constructed in somewhat general and simplistic terms, and thirdly it would fail to respond to the changing contours of that context over the past thirty years, changes that to some extent take an early cue from Feyerabend. Therefore I submit that an altogether more exciting and illuminating challenge to Torrance's theological science, comes from Feyerabend. Thus we cannot simply mount a critique or defence of Torrance's work in the form in which it appeared – for he like Feyerabend is writing in the context outlined above – rather we shall need to recalibrate his artillery to aim it towards Feyerabend's anarchic ground, thereby also directing it towards some of the standards of post-modernism. A profitable challenge test of any defensive system is the Maginot-line one – how does it stand up to an assault from an unexpected direction, particularly one which has gathered strength in recent decades?

For if Torrance depends heavily on Polanyi, how does he then fare against the challenge test of Feyerabend, whose basic take on Polanyi is that he, amongst others, has gathered some vital insights from within the philosophy of science, but has shied away from seeing them through

[97] See Preston in Preston, Munevar and Lamb (eds.), 2000, p. 83ff, particularly with regard to Feyerabend's last book *Conquest of Abundance*.

to the radical conclusions which they demand, instead seeking to compromise with a scientific establishment intent on defending its own interests?[98] Can Torrance defend himself from the call towards such radicalisation, or is it simply cowardice which leads him to the negotiating table? Whilst there are differing viewpoints within the philosophy of science, varying in hostility or friendliness towards theology, in allying himself with an 'establishment' figure (according to Feyerabend's estimate of him) such as Polanyi, any substantive challenge to Torrance's thought is liable to come either, to use a political analogy, from the left or from the right. There is little within Torrance's view of science which cannot be found in 'mainstream' philosophers of science.

So whilst Feyerabend cannot challenge Torrance on behalf of the 'central' consensus within the philosophy of science, he nonetheless poses a radical challenge from the left, which obtains real purchase on theological science, and though anarchic and deconstructive by nature, deserves to be taken seriously, not least for patterns of thought which it subsequently spawns elsewhere. For in challenging from the left, he lucidly encapsulates much of the post-modernism suspicion of the meta-narrative of Western natural science, even if he is isolated in presenting that critique from within the philosophy of science community. The challenge from the right – from more naturalistic and scientistic conceptions of science – is by no means without interest – witness the popular appeal of Richard Dawkins[99] and his 'new atheists'[100] – but within academic circles has been on the wane for well over half a century. No doubt many readers would expect a book such as this to deal with such a challenge. However it comes so much from the opposite end of the spectrum from Feyerabend, is frankly in scholarly circles rather 'old-hat', and is adequately dealt with elsewhere,[101] that such a space-demanding diversion must be declined. A defence of that flank would also be of less interest today, because Torrance's canons are

[98] Feyerabend, SFS p. 99 & AM, p. 295.

[99] Polkinghorne, 1998, p. 77, on why this approach is rather tired and simplistic.

[100] Beattie, 2007.

[101] For instance: K. Ward, *The God conclusion: God and the Western philosophical tradition*, DLT, London, 2009; *Why there is almost certainly a God: doubting Dawkins*, Lion Hudson, Oxford, 2008; A. McGrath, *The Dawkins Delusion? Atheist fundamentalism and the denial of the divine*, SPCK, London, 2009; A. Wilson, *Deluded by Dawkins, A Christian response to the God delusion*, Kingsway, Eastbourne, 2007; J. F. Haught, *God and the new atheists, A critical response to Dawkins*, Harris, and Hitchens, Westminster John Knox Press, London, 2008.

already facing pretty much in that direction, and also it is a direction in which the more thoughtful enemies of theology – despite the noisy few – are now much less sure of their ground than they were forty or fifty years ago.

Whilst every writer is framed by their own horizons and context, and whilst we must therefore repudiate with Gadamer the Enlightenment 'prejudice against prejudgments',[102] like Tracy we also admit that our distrust of this principle must not go as far as denying the potency of those thinkers able to transcend their own contexts. For even if the Archimedean viewpoint is physically an impossible one, our historical consciousness does allow us to identify works (classics) which have been able to speak to a wide variety of situations, so that the infinite reach of the Archimedean lever can be approached by imaginative endeavour, as mathematical series approach the limit of an infinite sequence. Thus whilst both Torrance and Feyerabend are responding no doubt to specific concerns and opponents – liberal and modernist Protestantism, rationalists (at least in the rather crude form of that term) and Popperians, respectively – a legitimate test is allowing each to face the other across impartial if not neutral territory.

Thus constructed our dialogue will therefore have a particular form. Both participants lay down a challenge to traditional Enlightenment notions of the 'scientific', as being too narrow and therefore false. Torrance tries to 'widen' and reconstruct these notions, taking promptings from the worlds both of natural and theological science, in order that the 'scientific' might ultimately be defined more fruitfully and properly. Feyerabend asks essentially 'why bother?' Why not deconstruct these culturally bound notions, and dispense with them? Why 'conquer such abundance'? Why not instead immerse, indulge and enjoy oneself within the great variety of human knowledge seeking traditions? Why stake out or cordon off a certain area of human endeavour with the label 'scientific'? If Feyerabend's 'knowledge without foundations' succeeds in dismantling Torrance's conception of science based around objectivity, then the question of the scientific status of theology is rendered meaningless, and Torrance's theological science project made irrelevant. Whilst space must be made for listening to them within their own disciplines, it is the significance of these

[102] Tracy in Küng and Tracy (eds.), 1989, p. 38.

thinkers with regard to more overarching epistemological concerns, which determines the meta-plain upon which they can meet.

To map out definitively that plain would be to offer a complete history of theology and natural science, so we shall not be able to examine every detail and will have to live with some degree of simplification. For instance in our account of the Enlightenment climate, we omit the writings of Kant's friend and rival Hamann,[103] within which we might discern the seeds of an already existing Enlightenment counter-culture. Thus our account deals with an admittedly brushstroke context – the seventeenth century scientific revolution, the eighteenth century Enlightenment, the rise of the historical consciousness in the nineteenth century,[104] the scientistic and positivist understanding of science which dominated the first half of the twentieth century, which was at least partly out-flanked by synchronic paradigm changes within natural science itself. This is adequate to identify within that context the epistemological features which both Torrance and Feyerabend felt compelled to wrestle with – Enlightenment foundationalism, a dualism of object and subject which hardened into objectivism, and a high and often ahistorical regard for the efficacy of human rationality.

The following analysis will show that their critiques of these features – though they also possess a degree of commonality – are divergent in their ultimate destinations. Our critique will find Feyerabend to be less careful[105] than Torrance, for in each case, he has a fourfold tendency towards crude binary deconstructions of accepted wisdom:

- Constructing a straw man of the position he wishes to defeat.[106]
- A paucity of historical examples which often prove nothing more than even the best rules have exceptions, and verges ironically upon a Popperian falsifiability approach which is allergic to anomalies.
- A failure to engage with any mediate positions – such as Torrance's – particularly when it comes to the meeting of 'traditions'.
- A dichotomous all-or-nothing belief that once you have rebutted a particular position you have therefore necessarily arrived at its radical opposite, whereas the philosophy of knowledge rarely works with such a binary dynamic.

[103] Griffith-Dickson, 1995, p. 240.
[104] Tracy in Küng and Tracy (eds.), 1989, p. 55.
[105] Preston, 1997, p. 209.
[106] Newton-Smith, 1990, p. 134.

However it is possible that these criticisms of Feyerabend are simply cover for the establishment defending its vested interests, and that the real fault lies in the timidity of those such as Polanyi and Torrance, who though they share Feyerabend's deconstructions when it suits them, are not sufficiently brave to see them through to their ultimate conclusions. Torrance's shoring up of epistemic defences with theological sandbags is then simply seen as the last ditch attempt of rationality to defend itself by an alliance with faith, the faded imperial power which had sometimes sided with the new empire and sometimes with its enemies. This then will be the pattern:

- Both Torrance and Feyerabend point to errors and shortcomings in enlightenment metaphysics.
- Torrance suggests a nuanced alternative, Feyerabend opts for more radical deconstruction.
- We must assess whether this divergence is due to Feyerabend's lack of care and rigour, and his rather flamboyant rhetoric, or whether it is due to Torrance's inertia and timidity.
- Often our assessment will hinge on whether we can allow Torrance the theological components of his construction without exterior support or verification.

It is an irony that the philosopher (Feyerabend) who wishes to promote human comfort over rational methods,[107] having failed to see that rationality is often fairly comfortable, leads us on therefore to pretty uncomfortable conclusions – though his riposte would be to ask 'comfortable for whom?' A key question will be: what does Feyerabend's attack on foundationalism actually entail? – is it simply timidity which leads Torrance to more moderate conclusions than Feyerabend, or is it in fact the case that Feyerabend's attack does not lead necessarily to his more radical conclusions, and that these are the products more of rhetoric than rigorous argument?[108]

One particular question mark must hang over Feyerabend's use of historical examples, and particularly his dependence upon Galileo. The lessons drawn from Galileo dominate the original text of *Against Method* to an almost absurd degree – the case study takes up chapters

[107] Newton-Smith, 1990, p. 145.
[108] For my answer see p. 227.

6–13 of (in its first edition) an eighteen chaptered book.[109] This is by no means untypical – the important principle of the indirect refutation of scientific theories is backed up with one rather dubious example – that of Brownian motion[110] – again ironically this principle is important in rebutting Popperian ideas over falsifiability!

Some commentators have reacted to Feyerabend's revisionist assessment of Galileo[111] by a straightforward defence of Galileo, but far more

[109] In a letter to Lakatos, Feyerabend reports the following:

"Today in a lecture I defended the Church's stand for law and order against the anarchist Galileo, and convinced everyone that Galileo was a bastard while Bellarmino was wise, scientific, humanitarian etc." (Feyerabend in Feyerabend & Lakatos, 1999, p. 333).

Lakatos in turn was forced to admit the truth of at least some of Feyerabend's case:

"According to Feyerabend *nothing more can be said*. Feyerabend's account is much more difficult to rebut than anybody else's. Indeed, we may in the end have to admit that Copernicus's and Kepler's and Galileo's adoption of the heliocentric theory and its victory is not rationally explicable, that it was largely a matter of taste, a *Gestalt-switch*, or a propaganda victory." (Lakatos, 1978, p. 178).

This is of course at odds with the 'traditional' portrait of Galileo, as the resilient champion of scientific truth in the teeth of fundamentalist religion. Instead, according to Feyerabend it was other less scientifically reputable qualities which secured Galileo his place in history: "Galileo's strong faith in Copernicus, his ebullience, his propagandistic ability, his willingness to cheat played a decisive role in the battle that was about to begin." (Feyerabend, KSR, p. 119).

Since coming across such revisionist views from the logical positivist Philipp Frank at Alpbach in the mid-1950s (See Feyerabend, AM, p. 257), when such a view was decidedly novel, Feyerabend returned repeatedly to Galileo as the paradigmatic scientific anarchist.

[110] Feyerabend, AM, p. 27.

[111] Feyerabend's revisionist views are predicating upon the understanding that Galileo's ideas contradicted the empirical evidence *as it was known at the time*. Furthermore heliocentric ideas were not new, but the resurrection of an old and convincingly discredited Pythagorean research programme. The fact that afterwards, sometimes centuries afterwards, more precise empirical support was produced for the heliocentric hypothesis, should not obscure the fact that judgements of empirical evidence have to be made in terms of the best available evidence at the time. Indeed there was scarcely a shred of evidence unambiguously in Galileo's favour. There was,

- The tower argument – a heavy object dropped from a tall straight tower, would be expected to fall some distance from its base if the earth was hurtling through space;
- Lack of stellar parallax;
- The planets did not seem to change their size and luminosity to the extent predicted;
- The more intuitive sense that we are not moving at thousands of miles an hour.

"Copernicus' system was far more complicated than the Ptolemaic one, involved many *ad hoc* assumptions, had mechanical difficulties, and introduced staggering assumptions about the distance of the fixed stars. Yet Copernicus held to his claim that his system had unique harmonies that proved to be real, even though he could only describe these harmonies in emotional passages." (Gelwick, 1977, p. 86).

damage has been done to Feyerabend's anarchic thesis by those who have happily adopted his views[112] on Galileo, but have declined to draw a link to Feyerabend's more general epistemological corollaries. The pragmatic rationalist is not concerned with the fact that, particularly in the past, not every successful scientist has proceeded according to the rules – after all it is entirely possible that luck and charm might play a part in scientific research[113] – all he needs to show is that scientists *fare better overall* if they act rationally and methodically, not that they do so in every case.

Others such as Feigl[114] and Reichenbach[115] have distinguished the context of discovery from the context of justification. In other words

---

Galileo met these problems with a variety of stratagems. He introduced new theories, such as the relativity of motion to explain i. His findings with the telescope, which were ambiguous (relying on the then highly contentious assumption that an instrument efficacious in the terrestrial field was therefore equally so in the celestial one), were used to counter iii. Evidence for ii would not be found for over two centuries. Curiously the one piece of empirical evidence that Galileo produced in his favour, based on his observation of water slopping around in Venetian barges, was the behaviour of the tides, which he put down to the motion of the earth. As we all now know, Galileo was wrong: the tides are caused by the gravitational pull of the moon. This is doubly ironic, as this idea, did *not* lie in the future, but was already common in Galileo's day, found as it was in Kepler. (Popper, 1979, p. 173).

Neither did the heliocentric theory, counter-inductive as it was, provide advantage in terms of scope. Indeed it profoundly diminished the more or less universal scope of the Thomist Aristotelian synthesis:

"Aristotelian dynamics was a general theory of change comprising locomotion, qualitative change, generation and corruption. Galileo's dynamics and its successors deal with locomotion only, other kinds of motion being pushed aside with a promissory note." (Feyerabend, AM, p. 121).

Given that the heliocentric theory contradicted Scripture, Aristotle and the empirical evidence, it is unsurprising that Galileo's ideas did not win immediate acceptance. Therefore Galileo adopted other less reputable methods, such as propaganda. His fall from grace was not simply the reaction of religious or scientific conservatism, but a judgement on his questionable personal behaviour, particularly within the confines of the conventions of patronage, or as Feyerabend puts it more bluntly he "had broken his promise and had tried to hide behind lies." (Feyerabend, FTR, p. 108).

As to the supposedly rough justice handed out to Galileo, one must note the rough and brutal character of the times, to see that his accusers actually proceeded with a degree of patience and leniency. Citing C. H. Lea, Feyerabend asserts that the notion of the 'Inquisition' as synonymous with inhuman judicial process is a great success for secular propaganda – by the standards of the day, trial and punishment by the Inquisition was preferable to submitting to similar procedures by the secular authorities. It is also the case that in an age when the eternal destination of one's soul was taken for granted as the overriding moral imperative, any idea which might put the well-being of others' souls at risk was considered with the utmost seriousness.

[112] See for example Sobel, 1999, Shea & Artigas, 2003.

[113] Bhaskar, 1991, p. 14f.

[114] Feyerabend, AM, p. 148f.

[115] Scheffler, 1982, p. 2.

whilst the initial discovery and promulgation of a scientific theory (the Kuhnian revolutionary stage) might feature large elements of chance and propaganda, the wider dissemination, reception, adoption and development of the theory (the normal stage) will be relatively free of such elements. Feyerabend's thesis is that by then it is too late, because once a theory has gained a certain degree of consensus, the "show has been rigged in its favour."[116] Therefore Feyerabend rejects the *quest* for normal science, as being simply a shoring up of vested interests and the status quo, and with the weapon of theoretic pluralism calls for constant revolution, in which theoretic diversity acts as a check and balance against the mythic pretensions of a particular scientific theory:

> Neither blatant internal inconsistencies, nor obvious lack of empirical content, nor massive conflict with experimental results should prevent us from retaining and elaborating a point of view that pleases us for some reason or another.[117]

He therefore also rejects any demarcation distinction between natural and theological (and other) sciences on the grounds (which for instance Kaufman gives[118]), that in natural science "*experimental* testing has become highly refined and reliable" – for Feyerabend that refinement is weighted heavily towards securing the 'right answer' for the reigning paradigm.

What is notable about this quotation from Feyerabend (no. 117) is not just its rather absurd force, but also its subjection of 'scientific' rigour to human flourishing[119] – whether "a point of view pleases us for some reason or another" is all important for Feyerabend, the voluntarist:

> The new ideal (of truth) should be simply to make people more pleasant and more interesting, to make life happier, to make the world better, to make the beer better and so on.[120]

To look forward briefly to our destination: our conclusion will therefore be that whilst Feyerabend's radical critique functions as a valuable challenge test to Torrance's own more measured and theologically

116 Feyerabend, SFS, p. 102.
117 Feyerabend, AM, p. 183.
118 Kaufman in Moore & Scott (eds.), 2007, p. 18.
119 See Preston in Preston, Munevar & Lamb (eds.), 2000, p. 91.
120 Feyerabend, *Science Freedom and Good Life* in The Philosophical Journal, 1968, p. 130.

secured critique by questioning whether Torrance has seen his insights through to a sufficiently radical deconstruction of the notion of 'the scientific', ultimately that challenge tests fails – though that is not to say that at no point does Feyerabend secure any purchase upon Torrance. Chiefly Torrance does not adequately engage with questions of the sociology of knowledge which play perhaps a more vital part in the development of Christian doctrine than in any other academic pursuit.[121] Here Feyerabend's work, and more influentially that of Kuhn, has demanded that every discipline, from physics to theology give an account of the historical, political and sociological factors which have affected or indeed infected the development of paradigms and theories. Unhelpfully for theology the traditional separation of systematic theology from church history has tended to overlook the interrelation of ideational and political / sociological factors which have influenced the development of ecclesial theology. After all though it may be historically doubtful, there can scarcely be a more flagrant example of the intrusion of political factors in scientific debate than the Eusebian tradition that Constantine himself suggested the use of the term *homoousios*.[122]

* * *

This is not a book concerning a certain area of systematic theology, although Torrance coming from a particular Christian theological tradition will require us to both work with, and sometimes critique his particular interpretation of a theological doctrine (e.g. revelation). The dialogue is primarily about epistemology and method rather than systematics *per se*. Our primary intention is therefore not to evaluate Torrance's theology as a whole or part, but to ask whether that theology coheres within his own conception of theology *as a science*, and more importantly whether Feyerabend succeeds in undermining either that conception or that coherence. That is not to say however that we can *start* by giving Torrance's understanding of 'science' prior to and independent of his theology, for his conception of science, of the structure of human understanding, is permeated with theological concerns. There are for Torrance not two moments to theological science – the theological and the scientific, they are interpenetrative concepts.

---

[121] Greinacher, in Küng and Tracy (eds.), 1989, p. 228.

[122] Need, p. 52, 2008; "Throughout the period [of the councils] theological problems were intertwined with political ambition", ibid, p. 110.

According to Torrance there can be no prolegomena in which we state what a science is, and then show that his theology fulfils the requirements.[123] Neither can we abstract from this interlaced vision a non-theological epistemology, for without particular theological commitments and therefore lacking key supports, the structure of Torrance's epistemology collapses – a concern for those prejudiced against theological commitments carried over into other disciplines.

Given their wide erudition, the holism of vision and variety of illustration employed by both Torrance and more particularly Feyerabend (part of his anarchism resides in his willingness to use examples and illustrations from anywhere he chooses), it is also important to point out that we shall have to handle a variety of topics – e.g. post-modernism, philosophy of language, hermeneutics, logic, scientific method – and it will be necessary to give some account of these as they occur in and impinge upon our dialogue partners. However, at no point should these things be understood as *the topic* of this book, as though a full account or assessment of post-modernism (or whatever) is being undertaken. The challenge test to Torrance is Feyerabend's, so that for instance with regard to realism, whilst we might seek to frame the debate within the present day discussion of this topic, we must also remember that Feyerabend's anti-realism knows no such sophistication, and consists of a rather inchoate and ambiguous reaction to his own former super-realism.

Indeed it is my contention that whereas Feyerabend suggested anarchy as a 'medicine'[124] in the days following the maladies of logical positivism and falsificationism, he would raise a wry posthumous smile at the way in which some of his ideas have been adopted by a whole tribe of social constructivists, which whilst vociferous and voluminous in certain disciplines, are of very limited value. So neither is it possible to offer an account of this industrial output, which regularly (for instance) takes Feyerabend's incommensurability thesis as the starting point[125] – that is to take Feyerabend too far, too seriously and too programmatically – his effervescent remedy is designed to deconstruct, not to form the starting point for a whole new edifice. He himself

---

123 Torrance, 1969, p. 10 – "It is scientifically false to start with epistemology" – see also p. 106 & Patterson, 1999, p. 46; Colyer, 2001a, p. 323, "Content and method are interrelated … ontology and epistemology ought to unfold together."

124 Feyerabend, AM, p. 17.

125 Hacking, 2000, p. 200.

despised the notion of belonging to, let alone founding, a 'school' of thought.[126] We must be bound to some extent, by the limitations imposed by choosing Feyerabend and Torrance as our dialogue partners.

Why then, as critical friend to a theologian who barely mentions him, take an anarchic philosopher of science who despite a certain 'solicitude for theology',[127] had in all likelihood never read any of Torrance's writings, let alone engaged in any form of substantive dialogue with him?

It would be possible to choose an alternative, less radical, philosopher of science – Polanyi is the obvious candidate – and appraise Torrance's project within their critical framework. An example of this approach is found in Weightman's 1994 *Theology in a Polanyian Universe: The Theology of Thomas Torrance.* However this would be a rather different exercise, assessing whether Torrance utilises the material of the philosophy of science establishment correctly and to good effect, rather than providing the sort of critical challenge which Feyerabend does.

Murphy's and Weightman's approaches underline the basic asymmetry between Torrance and Feyerabend which proscribes such an approach in this case. For in both those cases a theological programme – Pannenberg / Torrance – is assessed within a constructive programme from the philosophy of science – Lakatos' research programmes / Polanyi's personal knowledge. However it is not possible to show that theological science meets Feyerabend's criteria for good science, because his approach depends upon deconstructing that very notion, as constituting a socially constructed defence of vested interests. Rather Torrance's theological science must be shown as either being undermined by or resilient to Feyerabend's anarchic challenge – arguing as I do, that despite its anarchic nature it is a challenge which is coherent enough to merit a response.

Finally, given the observation that Feyerabend was not a terribly careful thinker[128] and somewhat prone to exaggeration,[129] it might well be asked why a more cogent and contemporary exponent of postmodern thought has not been selected? Whilst not impossible, it is not

[126] Feyerabend, TDK, p. 123.
[127] Watkins in Preston, Munévar, Lamb (eds.), 2000, p. 56.
[128] Preston, 1997, p. 209.
[129] Achinstein in Preston, Munévar, Lamb (eds.), 2000, p. 37.

attempted here because Feyerabend remains more or less alone, and perhaps indeed isolated, in offering a post-modern perspective *as a philosopher of science*. Philosophers of science on the whole are more concerned with responding to the challenges of a movement liable to question the over-arching ambitions of the scientific meta-narrative, rather than picking up those challenges and becoming the 'enemy' within. As we have also hinted, it would further widen the scope of this book and with many post-modern writers necessitate further detours into grounds foreign to Torrance, such as literary theory.

CHAPTER THREE

TORRANCE: THEOLOGY COHABITING WITH NATURAL SCIENCE

Although Torrance insists that he is simply travelling further along the road in the direction in which Barth pointed (see footnote 36, chapter one), it is nonetheless the case that Torrance travels so far down that road, constantly interrelating and cohering theological and natural science into a unitary framework of knowledge, that he has left Barth so far behind as to lose sight of him, not least because it forced him to reassess Barth's ban on natural theology.

This balance between maintaining the Barthian integrity of theological science, and offering appropriate bridges to other disciplines is summed up neatly in one sentence by Carvin[1]: "Religion must learn to live with whatever cosmology, whatever theory science provides, but on no account must it ever marry any of them." Polkinghorne offers a similar distinction between 'assimilationists' (the marrying kind) and 'consonantists' (cohabiters), classing himself as the latter.[2] The reason that it cannot marry them is that otherwise (as Bultmann argued) when scientific paradigms change (dramatically – as they have and no doubt will do) so the fundamental beliefs of the Christian faith would also be threatened.[3] Galileo's heliocentric theory appeared as such a threat because it threatened *both* the Bible and Aristotle thus undermining both poles of the Thomist synthesis.

[1] 1983, p. 306, see also McGrath, 2002, p. 134 for a similar expression: 'the folly of hitching', and Forster in Torrance (ed.), 1980, p. 108.

[2] Polkinghorne, 1998, p. 86.

[3] See McGrath. Vol. 3, 2003, p. 220 – McGrath makes a good point against Bultmann, in pointing out that Christianity had already passed through other cosmological paradigm shifts – in particular the medieval one to the Ptolemaic system – and that adjustment and accommodation rather than radical demythologising was required. For as Barth had concluded:
"The Christian Confession has in course of the centuries passed through more then one world-picture. And its representatives were always ill-advised when they believed that this or that world-picture was an adequate expression for what the Church, apart from creation, has to think. Christian faith is fundamentally free in regard to all world-pictures, that is to all attempts to regard what exists by the measure and with the means of the dominant science of the time." (Barth, 1949, p. 59).
Also Alston, 1991, p. 242 & John Paul II, p. 75, 1998.

Theology must learn to 'live with' whatever culture it finds itself in, firstly because it has no option, but also because it is religiously undesirable for Christian believers to retreat into a self-referential sect. Theology cannot therefore be independent of, nor dependent on, a certain natural scientific paradigm. There can be no retreat into either the polarities of fundamentalism or deism seeking hermetically to separate theological and natural truths. Neither should the necessary yet ultimately tentative and contingent synthesis of theology with the natural scientific paradigm of the day, whether that paradigm be Ptolemaic, Aristotelian, Newtonian, entail that a rebuttal of the latter would fatally undermine the truth claims of the former. I think it is worth quoting Torrance at length here, particularly as it throws light onto his relationship with his theological peers:

> Now there are people today who insist that the advances of modern science in the understanding of the universe make it necessary for us to discard the fundamental framework of the Christian faith as one that is bound up with an obsolete cosmology. Even the basic concepts of the classical creeds have to be changed for they are alleged to be but constructs of an objectifying kind of thought that makes use of transitory patterns in man's understanding of the cosmos in order to establish itself. Hence we must dispense with the objective forms in which faith is cast and find new conceptual forms congenial to modern man in which to express our attitude to the universe, and so on.
>
> If this is what is called the 'new Reformation' then it is a disastrous misunderstanding. Certainly the church can never stand aside from what is going on in the world, for it is only within the world, and not outside of it, that she lives and acts and speaks, and fulfils our mission. But let us see what actually happened in the eras of great cosmological change in the past. In each of them theological foundations underwent their due measure of adaptation and restatement, yet in each the essential imagery and the basic conceptuality of Christian doctrine did not change and the church adjusted itself remarkably well. There were certainly great struggles, but the apostolic and catholic faith proved adequate for all that was required of it in the change and advance in the understanding of the universe.[4]

This is indicative of Torrance's status as a 'type four' theologian according to Frei's typology, substantiated by David Ford in *The Modern Theologians*[5] (though confusingly and inexplicably inverting Frei's nomenclature). This typology forms a sliding scale of how far a

[4] Torrance, 1965, p. 260.
[5] Ford in Ford (ed.), 2005, p. 2.

particular theology has accommodated itself to the 'modern secular philosophy or worldview' of its time. Ford's description of type four is as follows:

> It insists that Christian identity is primary and that all other reality needs to be construed into relation with it, but also that Christianity itself needs continually to be rethought and that theology must engage seriously with the modern world in its quest for understanding,[6]

In the list of type fours we find Barth, Anselm, and Athanasius (fitting Torrance's self-description perfectly[7]).[8]

The second paragraph of the above quotation (number 4) indicates how and why Christian theology must learn to live with the natural science of its day (thus avoiding type five), whilst the first paragraph indicates the ways in which, according to Torrance, type two and three theologians have misunderstood the theological implications of the findings of natural science. Torrance sees this tendency towards over-accommodation as a constant danger in the history of Christian theology, and he identifies Gnosticism,[9] Arianism, scholasticism[10] (both Roman and Protestant[11]), and modernist demythologizing[12] as four points at which theology has been tempted to yield too much to (natural) philosophy – thus accounting for the diversity of theological paradigms and 'patterns of church order', despite the unity of its object.[13] A Christian metaphysics is possible, but it can only be built up from a truly Biblical theology, and where it incorporates elements from the current paradigms of natural science, it must hold them contingently. As Barth saw, this is not only advantageous for the integrity of theology, but also for that of natural science, which must free its research from theological repercussions.

---

[6] Ford (ed.), 2005, p. 2, though of course with regard to the Patristic period it is to be understood that Christian identity is encountered at a more formative stage, and that identity is more that which is in question.

[7] Torrance, 1990, p. 68.

[8] See discussion at 'Fluid axioms', p. 227; Torrance, 1995, p. 425.

[9] Torrance, 1995, p. 106–7.

[10] McGrath, 2003, Vol. 3, p. 278.

[11] Heron in Colyer, 2001.

[12] Torrance, 1971, p. 79; see also Osborne, 2006, p. 367 & p. 180 – here Osborne notes the longevity of the effects of such interpretative frameworks – here accusing Bultmann of using post-apostolic Gnostic categories to determine the interpretation of Biblical texts.

[13] Torrance, 1985, p. 119.

Although fundamentally fruitful and true, and again to offer premature conclusions, the difficulty with Torrance's approach is that one must state at the outset the contingency of natural science and its relation to theology, lest natural science compromises theological identity. However *thenceforward* theologians must proceed essentially forgetful of this fact – otherwise there is the opposite danger that natural science will lose its integrity to theology (Torrance, in foregoing Barthian caution, comes close to this, in often seeming to favour particular interpretations of physical theory, such as the Copenhagen interpretation of quantum theory,[14] *for theological reasons*). Theological and natural science must proceed in the parallel awareness that both are contingent and fallible, that each must pay heed to what the other is saying, and that both forfeit their scientific identity if they seek to dictate to the other methods or results pertaining to the other's particular domain of study. It is always difficult to keep these in tension, especially as it will be tedious to keep repeating them – thus sometimes theologians will have to proceed *just as if* the scientific paradigm of the day is infallible, but with fingers crossed behind their backs.

Torrance sounds therefore at times, as he writes about the consonant framework of theology and natural science, *as if* we have finally reached an omega point of co-existence, so that on the surface it *looks like* the Thomist synthesis or the deist treaty, of which Torrance is so critical. Weightman's charge: "No less than Rudolf Bultmann (or Thomas Aquinas), Torrance has chosen to appropriate philosophy to help illuminate Christian faith,"[15] is therefore justified in terms of appearance, but not in terms of stated method. Whether Torrance is therefore wholly consistent with that stated method is in doubt, as we shall conclude. Neither is Barth's coyness an answer, as ultimately it too can be construed simply as non-engagement.

Torrance is able to maintain that he stands in the tradition of Barth, despite his un-Barthian end-point, precisely because, rather than it being his theology which has taken an about turn, *it is natural science which has evolved* in a certain direction in the twentieth century, not only in its paradigmatic shifts of content (particularly within physics) but also more importantly with regard to the epistemological and ontological outlooks implicit therein. Thus his engagement with

[14] Torrance, 1969, p. xvi.
[15] Weightman, 1995, p. 275.

natural science extends as far as grappling with these paradigm changes, in order to distil from them results and changes of outlook which are significant for theology at the meta-level of epistemology. The three major changes which he focuses upon are relativity and quantum theory within physics, and the Gödelian theorems in mathematics. Whether theology can now live with natural science as peaceable and fruitfully as Torrance contends, is based in no small measure of whether his assessment of these changes is correct.

The theory of general relativity

Torrance's favourite and most successful example of a development in twentieth century science which has positive implications for theology is Einstein's development of relativity theory – in particular general relativity theory.

Fundamental to Kant's philosophical project in the *Critique of Pure Reason* was the *a priori* apprehension and intuition of synthetic truth, the most enduring example being the Euclidean nature of absolute time and space.[16] Without any experience of it, that Euclidean nature could be grasped and derived. Newton had been unsettled by this prior commitment, but most classical science proceeded with it as an unquestioned working assumption.[17] The questionable component of this intuition had long been identified as Euclid's fifth parallel postulate, which unlike the other four had always seemed somewhat less than self-evident, and by 1763 there were 28 'proofs', all flawed, of the fifth from the other four.

Implicit in the project of general (though not special) relativity theory[18] is the development, by mathematicians such as Reimann and Lorentz, of non-Euclidean geometries. By employing such geometries, relativity theory was able to predict empirical effects which were beyond the scope of classical mechanics. Thus natural science was left with a choice between abandoning the idea of the Kantian *a priori* or

[16] Kant, 1998, p. 162.

[17] Frege for instance thought that non-Euclidean geometries were merely interesting mathematical puzzles, with no bearing on reality, which could only ever live on as 'historical curiosities'; Frege, 1997, p. 252.

[18] Special relativity theory, whilst not wholly lacking non-Newtonian / Kantian properties such as time dilation, employs a four dimensional Minkowski space which is very similar to Euclidean space – the Lorentzian metrics employed by general relativity theory are radically non-Euclidean.

abandoning the theory which best explained various features of the physical universe – scientists generally chose the latter, if only because they were generally unaware of the philosophical implications of their choice.[19]

Thus physical science took a lurch towards the *a posteriori*, congruent to the demands of Torrance and Barth that (in their case theological) science should be *nachdenken*,[20] a demand which also features strongly in the work of McGrath.[21]

The effects of the Kantian *a priori*, which Einstein[22] and Popper[23] both categorically rejected, were twofold within theological science, one in terms of method and one in terms of content. In terms of method, Barth[24] (though he dates the change earlier than Kant, there is no doubt that he implicates Kant in consolidating this change) thought that after Kant, the category of the synthetic *a priori* had come to function as a universal prolegomena to all sciences, in that all scientific findings posterior to experience were to be framed within that prior intuition. Whereas the category of the synthetic *a priori* had proved vital for those who wished to develop theology and the other sciences *more geometrica* – foundationally from the base of fixed axioms, using only deductive inferences, well-defined concepts and terms, and inductively verified empirical results – this project, which Torrance and Barth saw as so stifling to theological realism and divine freedom, had been undermined from the inside, not by theological prejudice, but by the very sciences which it upheld as paradigmatic. For Torrance, Einstein's merit was not as a technician of normal science,[25] but as a Feyerabendian innovator, who had broken through an alliance of philosophy and natural science, an alliance which had imprisoned theology in a deist gaol.

After general relativity theory all of the system was brought within the orbit of *a posteriori* examination – 'axioms' had become fluid and revisable, and the rigidity of the system had suddenly become far more flexible. All sciences according to Kant had, to use a modern phrase, to 'think inside the box' – an apposite phrase, because the Kantian *a priori*

---

19 E.g. Popper, 1979, p. 129 or Hacking, 2000, p. 45.
20 Barth, CD, II/1, p. 63.
21 McGrath, 2003, Vol. 3, p. 278; McGrath, 2002, Vol. 2, p. 236.
22 Einstein, 1934 p. 63.
23 Popper, 1992, p. 29.
24 Barth, CD I/2, p. 4.
25 Torrance, 1965, p. 73.

intuition of time and space, posited time and space as an ideal and prior 'box' into which all our thinking about the material world must fit.[26] This Kantian notion had effects on the content of theological science, for it tended to seal off this box from divine action, and operate with a deistic disjunction between God and 'the box.' For instance, Torrance believes that the Lutheran and Calvinist controversy over the *extra Calvinisticum* is predicated upon the former's container view of space-time.[27] Torrance saw the effects of this Kantian legacy as deeply embedded within the Protestant theology which followed it.

> So long as people operate with an axiomatic disjunction between a noumenal realm of ideas and phenomenal realm of events, nothing more than a merely moral or poetic or symbolic or mythological meaning can be given to the Biblical account of the saving interaction of God with us in the world of space and time, or to the Christian message of the incarnation, crucifixion and resurrection of Jesus Christ.[28]

Two (admittedly slim) of Torrance's volumes are devoted to the implications of this particular view of space-time on two central doctrines of the Christian faith: *Space, Time and Incarnation, (1969a),* and *Space, Time and Resurrection (1976).* Torrance believed that these dualisms, given new life within Western culture by Kantianism and renewed potency by the "cross fertilisation of English Deism" with this "German idealist tradition,"[29] had lain embedded within Hellenistic philosophy, and through that source had permeated into varied cultural and academic traditions. He goes on controversially to maintain that some of the Fathers – Cyril, Irenaeus, paradigmatically Athanasius – had resisted such Hellenisation, in a way in which Protestant thought had largely been powerless to exempt itself.[30]

Torrance is of the view that it was Kant who, in reflecting on the Newtonian system, forced it into a rigid and mechanistic framework (after all Newton himself had thought that various astronomical anomalies were caused by God himself restoring the planets to their designated paths[31]), whereas McGrath tends to the view that:

> The sciences which Kant knew were Euclidean geometry and Newtonian physics. Given that Kant's philosophical reflections were partly shaped

[26] Kant, 1998, p. 162; Einstein, 1993, p. 157, rejects the idea that such a 'box' exists.
[27] Torrance, 1971, p. 126.
[28] Torrance, 2001, p. 27.
[29] Torrance, 1971, p. 108.
[30] Torrance, 1995, p. 228.
[31] Lakatos, 1978, p. 217.

> by these outdated sciences it is difficult to avoid the conclusion that the validity of his philosophy in general must be questioned.[32]

It is therefore undecided whether it was Kant who was guilty of forcing Newtonian mechanics into a rigid framework foreign to it, or whether the system itself at least suggested such a framework. In whatever direction the relation lies, the point remains, that the alliance of natural science and philosophy forced eighteenth and nineteenth century Protestant theology in a certain direction, which was eventually shown, in all three disciplines, to have been a dead end.

Torrance also makes a strong analogy between the relationship between geometry and cosmology on the one hand, and natural and revealed theology on the other. Torrance interpreted Barth's ban on natural theology as being based on natural theology's prior and independent nature, parallel to how Euclid's geometry was, in the Kantian scheme, prior to and independent of empirical cosmology. However just as geometry is formulated, in the relativistic scheme, posterior to and contingent upon an encounter with the material cosmos, so natural theology might be restored to a position posterior to and derivative of revealed theology.[33]

In terms of content, Torrance notes that in general relativity theory the inertial frame of the observer must be specified. Thus the ideal impersonal observer of the Newtonian world (the so called 'God's eye' view) is turned into the embodied observer of the relative world, whose status must be detailed. In a similar vein Feyerabend points out that various quantities – such as mass and frequency – which were treated as absolutes within classical mechanics now become relative to a certain inertial frame.[34] This is important for his incommensurability thesis – the relationship between classical and relativistic mechanics being one of his primary examples of this thesis.[35] However I would contest the notion that such changes amount to the *logical disjunction* which Feyerabend requires for incommensurability[36] as we shall in 'The incommensurability thesis.' (p. 138)

Similarly we should not get too carried away with Torrance. Whilst relativity theory does require a more detailed specification of the

[32] McGrath, 2001, Vol. 1, p. 31.
[33] Torrance, 1985, p. 39.
[34] Penrose, 1989, p. 282.
[35] Although he admits to a degree of subtlety in Feyerabend's arguments, Hacking initially characterises such discussion as 'stupid' and 'boring' – Hacking, 1975, p. 115 & 116.
[36] See also McGrath, 2002, p. 162.

observer, it is sometimes forgotten that the classical reduction of the observer to a mathematical point was only ever a mathematical idealisation. For whilst we might model (for instance) the pull of the earth's gravitational field on us, as a point at the earth's centre of gravity attracting another point mass at our centre of gravity, this was only ever a mathematical technique, which ignores empirically negligible effects – disregarding the effect of third or more bodies, and regarding the pull of ourselves on the earth as zero – ontologically the theory acknowledges that we are not point masses, it is just that empirically the cost of that idealisation is negligible.

Furthermore whilst philosophers may insist that theories state in advance any assumptions (e.g. 'assuming the effects of friction to be negligible …'), often they are omitted in a brief statement of the theory designed for working scientists. Unfamiliarity can then make it appear as though disjunctions exist, whereas they are merely given that appearance for the sake of a piece of workaday scientific shorthand. For instance, when Popper points out the logical contradiction between Newton on the one hand, and Galileo and Kepler on the other,[37] the reason that those 'contradictions' are not seen as crippling to the coherence of classical mechanics (in which they are taught as coherent, or more strongly when the former is derived from the latter) is that they are caused by simple and divergent initial mathematical assumptions. It is true to say that the ease with which such mathematical idealisation can be pursued is made a degree more difficult within relativity theory (even more so in quantum theory), so that enlightenment ideals of objectivity, the ideally minimal observer 'God's-eye view', are more complex to formulate, but one should not get too carried away ontologically. Relations have risen up the ontological league table over against objects, but relations were never entirely absent in classical natural science.[38]

Furthermore, there is sometimes a tendency for theologians to view, in a rather imprecise manner, relativistic physics as somehow more open and less determinate, and therefore somehow more convivial to theology than the mechanistic world view of classical physics. Fields and relations somehow sit more comfortably with theological language than objects, momentum and forces.[39] However this is to mistake the

[37] Popper, 1979, p. 198; Kepler's laws assume that one body (the 'sun' or star) is far more massive than the orbiting planets.

[38] Popora in Archer et al. (eds.), 1988, p. 350.

[39] Jammer, 1999, p. 210f.

language game – when relativity theory talks about field, relations, continuums etc., it is talking about them within the language of physics, where they have mathematically definite significance quite different from when we enter a personal or social language game. In particular one should note that 'relativity' in physics, has nothing at all to do with 'relativism' in philosophy. So one needs to ask more precisely why relativity theory is any more conducive to theology than classical physics? It is no less deterministic (though quantum theory is), it simply has more variables, more initial conditions, and its use of partial rather than linear differential equations means that the system has a greater number of degrees of freedom than the classical one. Torrance is here correctly brought up by Luoma – indeterminism is not a mathematical corollary of field theory.[40] It is also the case that fewer people understand its complexities and can therefore challenge theology from the ground of physics. It has singularities (such as black-holes); rather strange and complex mathematical formulations; it is often counter-intuitive, but it is sometimes not clear exactly what all this gains for the theologian.

Big-bang theory might incorporate the idea of a 'creation' moment – but steady state theory was dominant well into the relativistic paradigm – the battle between an eternal universe and a chronologically finite one has run over several changes of scientific paradigm, and Christian theology has already responded to that debate in the past, notably with John Philoponus.

Relativity theory does not seem to be relevant to the question of human and moral freedom (again unlike quantum physics). Newton found a place for the action of God in the world, outside of his closed system,[41] and relativity theory does not seem to open up any new avenues in which God's action in the world might be known. Given that what we now consider the natural order actually contains within it disordered and ruptured elements of sin and evil, so that God's intervention should not be seen as a further disordering, but as a reclaiming reordering,[42] the miraculous is always going to be identified by its deviation from the natural (at least of our human perspective), and therefore any explanation which seeks to offer an explanation of the event

[40] Luoma, 2002, p. 108.
[41] Torrance, 1996, p. 226.
[42] Torrance, 1996, p. 213.

within the natural laws of the present paradigm is liable to reductive rationalisation.[43]

Whilst Torrance might find the parallel between Barth's rejection of dualism and Einstein's rejection of classical particle and field dualism (though still present in quantum theory) illuminating and consonant, no assimilatory inter-dependence can be established above and beyond a conceptual resemblance.[44] Whilst Torrance is an enthusiastic consonantist, in concentrating on the level of structural epistemological parallels he just about avoids the assimilationist tendencies of less careful commentators. Torrance's position must be made clear: it so happens that at this point in time, in fidelity to the reality and nature of their proper objects, natural science and theology are having to work with 'parallel methods,'[45] methods which of necessity mark a break with their recent past. Therefore natural scientists are likely to be more open to theological patterns of thought, and in a world impressed with their work, contribute to a more theologically conducive cultural milieu, so that we can allow free rein to our consonantist instincts. However whilst Torrance may understand these developments as a successful penetration into a reality created and ordered by God, he also must allow that this is a temporary marriage of convenience, and that should, for whatever reason, natural science pursue a divergent course in the future, neither theology nor natural science can invoke binding commitments, and with a degree of regret perhaps, they must pursue their own endeavours with a degree of separation and if circumstance dictates a renewed state of antagonism.

Quantum physics

Quantum mechanics also marks a logical and conceptual departure from classical physics. Like relativity theory it needs complex mathematical techniques and counter-intuitive leaps to understand. In particular it has two very odd philosophical features, which are interpreted in divergent ways within and without the natural scientific community, and have each given rise to much speculation within the theological community. Those two features are the indeterminacy principle and the measurement problem.

---

[43] Polanyi, 1998, p. 284.

[44] Torrance, 1984, p. viii.

[45] Torrance, 1969, p. 8.

Einstein was famously ambiguous about the indeterminate nature of the theory, even if later on he was reconciled to it.[46] When he uttered his famous aphorism to Bohr, 'God does not play dice with the universe', Bohr supposedly replied 'don't tell God what to do'.[47] For Heisenberg had shown that the precise measurement of the position of a quantum particle, necessarily implied a more than zero value of the uncertainty of its momentum, and vice versa. In fact the product of the two uncertainties can be shown to be greater than $h/2\pi$ (where h is Planck's constant).

The measurement problem showed that some properties of quantum particles which we would expect to be intrinsic to that particle did not seem to 'exist', unless that measurement was made. The most infamously gratuitous illustration of this problem is the cruel state of suspended animation endured by Schrödinger's cat.

It must be understood that both these problems are not to do with experimental limits on the precision of measurement, or any limit in human physical or mental abilities, but seem to be limits inherent within the quantum system itself – and therefore if taken in a 'realist' sense, reality itself possesses these counter-intuitive features. The Copenhagen interpretation takes these uncertainties at face value.[48] However other experimental results such as the E(instein) P(odolski)R(osen) effect cast doubt on this instrumentalist interpretation.[49]

These two features are particular to quantum theory, and mark a logical disjunction between quantum theory and both relativistic and classical mechanics.[50]

Whilst many scientists simply live with this disjunction at least for the present, there are those, such as Feyerabend, Penrose and Popper, who therefore view quantum theory as a staging post towards a more complete theory, albeit an historically persistent one.[51] Feyerabend's early defence of quantum instrumentalism[52] is immune to any criticism of it as a defence of global instrumentalism, in fact it is the opposite, being rather a critique of quantum theory. This defence of *local*

---

[46] Einstein in Maxwell, 1982, p. 32.
[47] See for instance Rae, 2004, p. 50.
[48] Rae, 2004, p. 54.
[49] Rae, 2004, p. 55.
[50] Newton-Smith, 1990, p. 228; see also Penrose, 1989, p. 5.
[51] Penrose, 1989, p. 452; Popper, 1992, p. 459.
[52] Preston, 1997, p. 67f.

instrumentalism is simply a 'tactic'[53] which in fact assumes a backdrop of global realism – and without such realism the incommensurability thesis will fail.[54] Feyerabend therefore insists that instrumentalism is not imposed *ad hoc* upon an unwilling realist quantum theory, but is built into its structure from the start. However this means that the early Feyerabend is guilty of passing off global realism as an assumption lying within the content of science, rather than a philosophical decision about the nature of science.[55]

Torrance endorses the Copenhagen interpretation of quantum theory,[56] not least because of the interpretation's philosophical implications, which are relevant to modern theology:

> one of the interesting results of this new positive and dynamic theology in the modern style is the parallel between its method and that which has emerged in quantum physics. This is particularly evident in the way that in which both physics and theology have had to treat the old antinomies between object and subject, thing and motion, being and act, determinism and freedom, etc.[57]

There are however other interpretations of quantum theory, which whilst they are not as widely used as the instrumental Copenhagen one, nonetheless command some support within the scientific community, have no crucial experimental or theoretical rebutters, and mark a genuine divergence in how this scientific theory is to be interpreted.[58] Torrance comes very near here to endorsing a particular

---

[53] The words 'tactic' should not be held to imply anything deeply strategic and programmatic in Feyerabend's development – it is simply a case of trying out various arguments over the years, and seeing where one ends up – quantum instrumentalism is one of Feyerabend's earliest topics (1958).

[54] Feyerabend, AM, p. 274ff.

[55] Preston, 1997, p. 73.

[56] Torrance, 1969, p. xvi.

[57] Torrance, 1969, p. 8.

[58] Firstly the 'hidden variables' interpretation (of de Broglie and Bohm for instance) is the most classically realist interpretation. It employs the same tactics as those of classical physics when it came across indeterminacy – assumed there was another variable, and searched until it was found. However the problem for this interpretation, is not only that there is as yet no sign of such a variable, but also that the hidden variable hypothesis introduces other disreputable features into the system (Rae, 2004, p. 33). It also seems to have been shown that no such variable could account for the quantum results so far obtained (Rae, 2004, p. 40). Whilst Van Fraassen flatly disagrees with this assertion, ("The 'no hidden variable' proofs as I have already mentioned, rest on various assumptions which may be denied." Van Fraassen, 1980, p. 68), Rae still concludes that "we should look for something other than local hidden-variable theories to resolve the puzzles of quantum physics." (Rae, 2004, p. 49).

interpretation of a physical theory *for theological reasons*, which the consonantist must be generally careful to avoid, if he has not made it absolutely explicit what he has done.

As McGrath[59] so rightly points out, the microscopic effects of quantum indeterminacy have spawned a tremendous amount of work which leaps unwarrantably to a frankly hideous defence of crude relativism, which whilst obscure and varied in many details, reveals nothing more than the basic scientific illiteracy of their authors, and is unworthy of engagement. Typical is Bryan Appelyard who claims that scientists have given up the notion of objectivity.[60] Whether it is fair for McGrath to blame this on (unnamed) "fashionable French intellectuals" is a moot point, particularly as McGrath's main target is Stanley Aronowitz, but certainly we make no excuse for ignoring these 'extensions' of the indeterminacy principle which no natural scientist would endorse.

Often, unlike Torrance, they do not have sufficient background or technical knowledge of the natural sciences, and sometimes seize on

---

Secondly another interpretation of the quantum theory is the 'many worlds' theory. This might have appealed to Feyerabend, firstly because it seems initially so unlikely, and secondly because it tenaciously revives, after two and a half millennia, a favourite theory of Democritus, which was sent into obscurity by the rationalist trio of Plato, Aristotle and later Aquinas. In 1957 Hugh Everett III postulated that quantum 'choices' were not really 'choices' at all, but each was fulfilled in a plethora of actual branching universes. The good thing about this theory is that it removes indeterminacy, and confirms experimental results. It is also consistent, because each branching universe, is thenceforward completely independent of other branches – and we can have no knowledge of the other branches (including the billions of different 'selves' which we have branched into). The problem with the interpretation is this 'extravagance of universes': "to postulate the existence of a near-infinite number of complete universes to resolve a subtle theoretical point seems to be introducing 'tigers' into our thinking with a vengeance" (Rae, 2004, p. 86). Recently the astronomer royal Martin Rees, amongst others (see *Multiverse or Universe*, CUP, Cambridge, 2007, ed. Bernard Carr), defended this 'multiverse' interpretation precisely to counter the theological suggestion that this universe was in any sense 'chosen'.

Thirdly, and most controversially, there is an idealist interpretation, verging towards mystical solipsism. Based around the quandary of the measurement problem, it seeks to take seriously this feature of the theory – that we 'make' quantum values by measuring them. The problems with this interpretation are several, and certainly seem out of proportion to the tiny experimental effects they seek to explain. Firstly what about quantum values which are not measured? Secondly what if two observers seek to measure the same value? (Rae, 2004, p. 71 & Rae, 2004, p. 147). Unless we are happy to inhabit some very unlikely philosophical territory, it is unlikely that this interpretation will yield any serious aid for natural science – not least because it calls into question the independent existence of nature and natural scientists!

[59] McGrath, 2002, Vol. 2, p. 283.

[60] Appelyard, 1994, p. 157f.

particular interpretations and features of a theory, thus binding their theology or philosophy to some rather unlikely scientific view point. When conjecturing from physical theories to more universal philosophical standpoints, theologians, Torrance included, need to say more succinctly what exactly they are trying to get out of those theories. Furthermore when theologians sign up to interpretations of theories which enjoy only limited support within the natural scientific community, they have to say what they think will be *lost*, if, as so often occurs, such minority views are rebutted or simply abandoned – is the discussion noting merely a passing resemblance, or is it committing the author to frameworks which it may prove difficult to subsequently unravel? I have already pointed out how this is important in relativity theory, and it is more important in relation to quantum mechanics.

Though the question of the existence of the human soul is,[61] since the publication of Popper and Eccles 1977 *The self and its brain*,[62] most often the principle focus of interest within quantum mechanics for theologians,[63] Torrance is more concerned with the sovereign freedom of divine rather than human action. Quantum indeterminacy could conceivably offer this freedom, by imagining God as somehow lining up almost countless quantum events in order to secure his will.[64] Such a solution is consistent with a God who is omniscient, omnipotent and omnipresent. However again whilst this has a certain epistemic neatness, it is difficult to see what other advantages it has over a more

[61] Rae, 2004, p. 147.

[62] Also see Popper, 1979, p. 232; 1969, p. 265.

[63] It acquires support from amongst others Roger Penrose. "I believe, that contrary to what appears to be the majority view, among physiologists, quantum phenomena *are* likely to be of importance in the operation of the brain." (Penrose, 1989, p. 194). Thus quantum indeterminacy provides the degree of freedom necessary to postulate a mechanism (the soul) by which quantum choices are lined up within the brain, placing the brain in the desired neurological state, resulting in macroscopic effects (quantum amplifiers) on the rest of the body and the exterior material world. Whilst this is epistemically a perhaps more comfortable solution than a Cartesian dualism (of extended and thinking substance) which postulates a substance of moral agency free (though operationally bound by locality and temporality) of the determinism of the material world, it is by no means invulnerable to other criticism. It is hardly a posterior theory based on evidence, but rather a bold conjecture to get out of an epistemic hole, vulnerable to a soul-of-the-gaps rebuttal should the content or interpretation of quantum theory change. Torrance does not discuss this theory in particular, though with his intense dislike of dualism, combined with the fact that, like most theologians, he is keen to conserve a human moral freedom, means that he refers more generally and approvingly to the 'open indeterminate non-mechanistic' systems of quantum mechanics.

[64] Ellis, Murphy, Robert Russell, & Thomas Tracey all follow Pollard in supporting this proposal – Peacocke and Polkinghorne are against – Southgate (ed.), 1990, p. 260.

traditional theological and physical picture of the universe. As we have already noted, the religious efficacy of miracle or sign is significantly devalued, if it can be reduced to an event which lies in conformity to natural possibilities – surely the point of the resurrection (for instance) is that it dramatically demonstrates God's sovereignty over the natural processes of death and corruption – *if Jesus is just very very lucky, what possible grounds can that be for Christian hope?*

As with the existence of the soul, this dependence of divine action on quantum choices is too much at risk of conjoining God's freedom to a particular and controversial interpretation of a physical theory, vulnerable to a future closing of the gaps, in which God's 'lever' in the world would be removed at a stroke. Instead we would do better to return to the meta-question, as Torrance would have it, of whether classical physics necessarily implies the theologically problematic 'deistic disjunction' between a mechanistic determined world and the noumenal realm of theo-logy, or whether other factors meant that this philosophical interpretation has been foisted upon a scientific theory which does not contain it as a result. The answer is ambiguous – logically no (Torrance is right – there is no entailment), but one can see why philosophy and theology were tempted to move, either consonantly or assimilatorily, in the direction they did. Barth here, in raising the problem before these developments in physical science, passes the test of prophetic insight, which is reduced in Torrance to the rather weaker gift of hindsight.

Torrance is more concerned with metaphysical issues within quantum theory which realign the relations of object and subject (a key part of his work – see chapter four), set limits of human knowing, and demand that even logic ('quantum logic'[65]) is justified posterially in the Lesbian manner. To do so demands that the indeterminacy of the Copenhagen interpretation must be taken in a realist sense,[66] as a fundamental uncertainty in the ontological depths of the material world, whereby in lurching towards the relational it becomes open to the divine.

There is however a lack of clarity about what exactly such a quantum logic might achieve, particularly in relation to traditional logic. Whilst Torrance might refer to traditional logic as overly 'narrow and restricted', it is difficult to be more specific about which parts of traditional logic need to be opened up and how – if we ignore the law

[65] Torrance, 1971, p. 7 & 12; 1969, p. 271f.
[66] Torrance, 1969, p. xvi.

of excluded middle at the quantum level (for example),[67] what criteria are we to give for its applicability? Or are we to give it up entirely? Torrance's concern with the metaphysical relations of logic and being obscures how those concerns might actually play out in descriptions of the material world. Physicists themselves (Feyerabend, with regard to this question, contrasts, with the utmost rudeness, the playfulness of Bohr and the seriousness of Popper[68]) have on the whole sat lighter to such questions, holding the quantum theory, as per the constructive empiricist Van Fraassen, to be empirically adequate, but leaving open its relation to 'reality' and indeed relativity theory. Natural scientists, generally a philosophically pragmatic tribe, intuitively and instrumentally answer with some ease the questions which cause so much angst for philosophers.[69] For fast or very large systems they use relativity theory, for nano systems they use quantum theory, and for medium paced, medium sized dry systems (i.e. the vast majority) they use classical mechanics.[70] In quantum experiments they are, in practice, able to distinguish without difficulty the system from the measuring apparatus. For the most part they operate with a hybrid of realist and instrumentalist interpretations – they ignore the logical inconsistencies of such an approach, because it gets results and predictions right, but they also retain a tacit belief that they are at least working towards something which is a description of reality at its deepest level. Feyerabend in his case for theoretical pluralism notes that such a variegated approach may in fact overthrow a more simple, complete and homogeneous (in this case Aristotelian) system[71] – and Weizsacker adds that this was also the case in the subsequent paradigm shift from classical to modern physics.[72] Philosophical purity is simply not a priority for working scientists.

Gödel's theorem

Frege was troubled by the fact that whilst many natural scientists are relatively unconcerned with the philosophical basis of their science,

[67] Torrance, 1971, p. 12.
[68] See Feyerabend, FTR, p. 185.
[69] Rae, 2004, p. 95.
[70] Feyerabend, CA, p. 141.
[71] Feyerabend, AM, p. 121.
[72] "A few decades ago physical science had a unified picture of the world. This picture offered a frame into which all known physical phenomena fitted … today this world no longer exists." Weizsacker, 1952, p. 11, opening words.

despite its supposed rigour, mathematics fares no better.[73] He then set about finding such a basis by attempting a synthesis between the results of mathematics and logic. Although it attracted initial interest,[74] a crushing blow was delivered to his project by Russell's paradox,[75] and he ceased work, suggesting in a late rather melancholic letter that geometry rather than logic might yet form a suitable basis.[76]

Russell made his own attempt to construct a complete mathematical system with Whitehead in his *Principia*. However this too was to come off the rails with Gödel's theorem of 1931, which proved that within any formal system, there must be undecidable true propositions. Given any finite number of initial axioms, the only way to prove some of the true propositions within the system is by axiomatic enrichment.[77]

Polanyi expounds the implications of the theory thus:

> Gödel's theorem has shown that the scope of mathematical formulae is indeterminate, in the sense that we cannot decide within a deductive system like arithmetic, whether any set of axioms comprising the system are consistent or mutually contradictory.[78]

Since this theorem, foundational questions of mathematics have been in abeyance, because the hope of a complete systemisation has not only proved to be elusive but impossible.

> The notion of mathematical truth goes beyond the whole concept of formalism. There is something 'God given' and absolute about mathematical truth. This is … mathematical Platonism, … any particular formal system has a provisional and 'man made' quality about it. Such systems indeed have very valuable roles to play in mathematical discussions, but they can supply only a partial (or approximate) guide to truth. Real mathematical truth goes beyond mere man-made construction.[79]

Gödel's theorem is a devastating blow[80] to mathematical formalism, which Torrance therefore explicitly ruled out.[81]

This quality of 'God-givenness' is grist to the mill of a positive theologian like Torrance. Speaking about the 'theological' qualities of awe and wonder he remarks:

---

[73] Frege, 1997, p. 200.
[74] Frege, 1997, p. 138.
[75] Waismann, 1965, p. 71.
[76] Frege, 1997, p. 303; also Dummett, 1978, p. 92.
[77] Torrance, 1996b, p. 276.
[78] Polanyi, 1998, p. 94.
[79] Penrose, 1989, p. 146.
[80] Penrose, 1989, p. 137.
[81] Torrance, 1980, p. 114.

> This is certainly the astonishing thing that keeps on striking the scientist with wonder and awe, as Einstein used to say, that there is already embedded in nature an inherent rationality which it is the task of science to bring to light and express.[82]

However he extends slightly recklessly the Gödelian insights to systems other than the mathematical.

> Gödelian theorems demonstrate that any logico-deductive system of arithmetic or logic of sufficient richness is consistent only if it is incomplete in itself, or if it is consistent, it is so only through reference to a level of rational connections beyond it – theorems that apply to mechanical as well as mathematical and logical systems.[83]

Others, in contrast, strike a more cautious note,

> However [with reference in this case to Penrose] Gödel's theorem is concerned with mathematical systems, and there are serious doubts as to whether extrapolations from it to other areas are legitimate.[84]

What Torrance grasps as being theologically important is the fact that no human epistemological domain, even mathematics, until then regarded as the paradigmatic fortress of necessary logical connections, could be seen as entirely internally self-sufficient and coherent, and therefore could not be a possible basis for any universal foundational system – even in mathematics, God's act of grace in creation had embedded a rationality which is ultimately beyond human formalistic procedures.

> A helpful and I believe legitimate extension of Gödel's undecidability theorem might be cast in the form: no syntactical system contains its own semantics, or, no dogmatic system carries within it its own truth-reference. To be consistent no logico-deductive or logico-syntactic system, at least of sufficient richness, can be complete, but at significant points must be open for reference beyond itself to a wider and higher system.[85]

The extension here to logico-syntactic systems is a dubious and rather unnecessary one since we have already noted Frege's failure to bind arithmetic to logic. If as we shall later discuss, theology, philosophy

[82] Torrance, 1971, p. 42 – Torrance, 1969, p. 122 & 184 describes Einstein's awe as "the pain and awe and excitement of radically new knowledge."
[83] Torrance, 2001, p. 70.
[84] Southgate (ed.), 1999, p. 179.
[85] Torrance, 1985, p. 123.

and natural science are all bound to ordinary language (heavily disliked by Frege) then Gödel's theorem is unnecessary to resist formalism within those domains. The overarching result that Torrance requires is that any system which has a 'relation to reality' must be an open one:

> Thus Gödel's discovery serves to reveal the limits of logical formalization or of axiomatization, since it shows that the structure of a formal system by its very nature must be open, and can therefore be free for new discoveries, while a closed system, one entirely consistent, and demonstrably so within itself, would not be true for it could have no relation to reality.[86]

Torrance is using the Gödel undecidability sledgehammer to crack the formalism nut. The theorem is a highly technical one, precisely specific to certain sorts of deductive mathematical systems. In that sense it therefore has no direct entailments within *any* physical or human science, for no such science is constructed solely in terms of such a system. Whilst sciences such as physics employ mathematical techniques to the point that mathematical manipulation comprises 90% of the work done, especially with the advent of super-computing, there are still always residual questions of applicability, quantifiability and empirical adequacy.[87] Thus there will always be, in whatever physical science, a component which Polanyi labels as tacit and personal. Conversely theology using minimal mathematical manipulation, is therefore even further from the sort of system to which the undecidability theorem would apply.

Just as with relativity theory and quantum theory, the question for Torrance is why does theology *need* such a result to maintain its integrity? It is not a closed formal system, but then neither is physics or sociology. Conversely even if Gödel had proved the opposite – that mathematics *was* capable of closure and consistency – that would not bind any other science, natural or theological. Again there is an epistemic 'parallelism' between theology and post-Gödel mathematics, but given the very specific scope of this theorem, there does seem little danger that anyone, even the severest Scholzian would expect theology to imitate the kind of formalism that logicians once attempted – it is indeed a case of sledgehammer and nut.

---

86 Torrance, 1969, p. 255.

87 No 'isomorphism' as Polkinghorne would have it (1991, p. 39f).

Conclusion

In his reflections on paradigm-shifting developments within natural science in the twentieth century, it is important to understand that Torrance grants to both theological and natural science an integrity and autonomy to determine the content, and indeed the methods, of their own science. Whilst the interrelations between their constitutive theoretical content must continually be forged within a 'renewed dynamic integration', one cannot dictate to the other with regard to content. If therefore there is a *conflict* at the level of *content*, it can only be resolved (or not) by further work in each discipline, and conversely any congruence or consonance will be contingent. Neither discipline can dictate content to the other, and neither should a particular theory or interpretation be favoured within one discipline because it accords with the wishes of another, though Torrance sails fairly close to the wind in favouring the Copenhagen interpretation within quantum mechanics.[88] Torrance is therefore not concerned with working out the theological implications of the content of natural science or the physical implications of theology, but rather in developing a metaphysical framework in which these two contingent disciplines might co-exist and interrelate. There is no rerun of former controversies such as that of Galileo or Darwin, which dealt with physical content, because when Torrance investigates contemporary natural science he is looking for principles at the meta-level of epistemology, rather than examining conflicts of content. There can be no question of a particular physical result being *necessary* for the possibility of theological science, though it is true that certain methods, conceptualities and theoretical frameworks are conducive to theological patterns of thinking, which in turn are not dictated by these, but correspond with the demand that theology must speak faithfully of God.

At this meta-level Torrance is however quite clear that twentieth century natural science has moved in a direction which is more conducive to and consonant with the methods of theological science, though if it had not, then one would simply have to stick at the point of Barthian remove, rather than try and forcibly engineer an alternative accommodation. According to Torrance therefore, errors have as a matter of fact been made at this meta-level in both directions: either contrasting

[88] He just about can get away with it, in that Copenhagen has 'majority' rule status within the natural scientific community.

the expressivist truths of theology with the 'objective' truths of natural science, or over-aligning them within the mis-relations of rationalism and fundamentalism, in which one party imposes unwarrantedly its framework, methods, and ultimately results on the other discipline.

Against the fundamentally hostile background of Kantianism *a priori* rationalism, mechanism, and objectivism, Torrance distils from modern paradigm shifts the following positive results:

- General relativity theory forces us:
firstly to dispense with the most important constrictions of the synthetic *a priori;*
secondly to revaluate the relation of object and subject;
thirdly to soften the hardened dualisms of field and particle;
- Quantum theory forces us:
firstly again to revaluate the relation of object and subject;
secondly, to question the frontiers of what it is possible for human beings to know, and underline their status as creatures reliant upon grace even in relation to intellectual objects;[89]
thirdly, reject any crude naturalistic determinism;
- The Gödelian theorems force us:
firstly again to revaluate the frontiers of the possibilities of human knowledge;
secondly to reassess and restrict the relation between the truth of statements and the possibility of their proof by logical means;
thirdly to deny logic and mathematics any absolute foundational and *a priori* privileges over other sciences.

The Ariadne's thread[90] for Torrance, is the battle between a dualism inherited from Greek philosophy which infects both classical natural science and theology, and the unitary outlook deriving from Hebraic thought which had to fight to assert its integrity during the Patristic period.[91] Torrance's need to develop a fairly simple set of antithesis here results in a rather crude handling of contested material.[92] Within natural science it eventually triumphs in modern physics which seeks to rediscover a unitary basis, especially in the thought

[89] Torrance is never afraid to use the epistemic capital inherent in Athanasius' dictum: "Thus far human knowledge goes. Here the cherubim spread the covering of their wings." Torrance, 1996, p. 193.
[90] Marley, 1992, p. 3.
[91] Torrance, 1985, p. 29; Colyer, 2001a, p. 57f.
[92] Luoma, 2002, p. 34.

of Einstein[93] (see p. 232f). For Christian theology therefore the battle has been to keep its nerve, and the results found in those such as Athanasius and Philoponus, who resisted Hellenistic dualist patterns of thought, which remained dominant and often unchallenged within natural science, right up until the twentieth century – only then was such Feyerabendian tenacity to be rewarded.

Torrance's unitary outlook therefore consists of a set of disciplines which have a certain family resemblance,[94] share certain 'con-sonances', but are also marked by a diversity proportionate to that of their objects. Thus any demarcation criteria of the 'scientific method', whilst possible, will substantially be imbedded within the disciplines themselves (see chapter four).

What Torrance is describing within natural science is what one might call a climate change. The enlightenment climate[95] was characterised by the importance of empirical adequacy (and demarcated what that amounted to), the aristocracy of reason (especially mathematical logico-deductive reason), the possibility of the *a priori*, and a linear progressive foundationalism favouring axiomatic systems. Theology was therefore liable to feel pressure either to adapt to the climate (by which theology was in danger of becoming non-theological), with a corresponding distortion of its integrity, or to maintain the primacy of revelation and theo-logic, and suffer a high degree of marginalisation (theology as non-science).[96] Polanyi's metaphor is neat and accurate:

> If divine revelation continues to be venerated, its functions – like those of the kings and Lords in England – are gradually reduced to that of being honoured on ceremonial occasions. All real power goes to the nominally Lower House of objectively demonstrable assertions.[97]

Smedes[98] refers to natural science's success as engendering an unwarranted 'scientistic ideology' within enlightenment culture. Barth had looked out at the weather and decided to stay indoors. Torrance's synthesis of natural and theological science, albeit tentative and contingent upon developments from 1900 onwards, must be understood as a reaction to such a scientistic ideology – as maintaining that such an

[93] Marley, 1992, p. 6.
[94] What Torrance calls their 'parallels of method' – 1969, p. 8.
[95] Dilthey, quoted in Griffith-Dickson, 1995, opening quotation.
[96] See chapter four: 'A new objectivity'.
[97] Polanyi, 1998, p. 266.
[98] Smedes, 2004, p. 9.

ideology was just that, an alien imposition upon science, rather than being a result or corollary of it. The defence of the previous Barthian era was to underline the fact that such an ideology is not scientific (i.e. it cannot be deduced from natural, or any other, science), but a naturalistic prejudice. The problem with showing that the climate has changed, and is now more favourable to theological science, is that whilst perfectly true, this does not undermine but may in fact be liable to reinforce such a scientistic ideology – if and when the weather turns inclement in the future, theology must recover this Barthian past, if it is not to once again be stranded, exposed to the elements.

Torrance therefore as we shall see, shares with Feyerabend, a critique of his metaphysical inheritance: empirical accuracy has been shown (by Feyerabend and others) to be tainted by theory; the search for universal reason, scientific method and criteria of scientific demarcation have all failed, and even our best physical theories are infected with anomaly and contradiction; the synthetic *a priori,* at least the Kantian examples of it, have been shown to be in fact posterior to experience; and all sciences have resisted strict axiomisation.[99]

So though this 'climate change' is more sympathetic towards theology at a meta-level, none of the changes in themselves are for Torrance necessary ones, they are renewed epistemic sympathies. Divine and human freedom can be defended without quantum indeterminacy. The quest for truth as enacted within theological and hermeneutical spirals can be defended without Gödel's theorem. The contrast between proper objectivity and objectivism can be shown without relativity theory. They are powerful allies, and make the task of the theologian that much easier in the teeth of a scientism which had assumed ideological pretensions. However the absence of bonds of necessity may act as a suitable restraint on those who would wish to rush the bride of natural science to the altar, rather than being happy to cohabit. Marriage in this case at least, may not make an honest woman of her.

In the subsequent and continuing climate change from modernism to post-modernism, (and as Kuhn and Feyerabend would point out such revolutionary changes are likely to be fairly piecemeal and fragmentary affairs) there has been an opening out epistemologically. This is Torrance's other problem – one can hardly say that these changes

[99] Indeed as we have seen Gödel has shown that such completeness is inherently unlikely – if it is impossible in mathematics, even if his results were thus restricted, it is unlikely it will occur elsewhere – apart from anything else science would have to dump mathematics.

have pointed decisively in the direction of Barthian theology, indeed to have done so would have compromised theology's integrity – Torrance is quite clear that the results, of for instance natural science, could not lead us directly to a particular theological destination (otherwise the way is open to a more substantive natural theology), but by their nature can only show us possibilities. Rather those changes have opened the field in a Feyerabendian way (though often without the irony) to a great diversity of paradigms and traditions, some consciously 'scientific', many others less so – for instance within medicine towards complementary and alternative techniques drawn from a breadth of cultural and metaphysical backgrounds. Torrance must not only show how this 'opening out' has released theology from the 'Babylonian captivity' of an enlightenment strait jacket,[100] but why its final destination is not a Feyerabendian 'anything goes'. The weakness for Torrance is therefore his undefended fideism[101] (albeit a 'defensible indefensibility'[102]), which may appear to post-modernism, as Pannenberg says, simply capricious.[103] However one should remember the Barthian point that philosophy cannot and must not fully justify theology, for then God's sovereignty would be made dependent upon grounds other than that of his self-revelation. For Torrance fideism is not an inconvenience or embarrassment, quite the opposite – it is a necessity.

What also counts against Torrance, is that the necessary tension in the relationship between theological and natural science is not fuelled by any identifiable negative developments – *the news is all good*, at least at the meta-level. A more firmly founded independence between these sciences would be secured if he were to elucidate any negative developments within recent paradigm changes. Instead one is left with the impression that *all* changes at the meta-level of natural scientific epistemology are conducive to theological science, and therefore the suspicion is that in fact theology is, despite the denials, being made simply to conform to the next paradigm shift just as surely as it was to the last. Whilst Weightman's charge[104] may be ultimately unfounded, it is not hard to see why the circumstantial evidence might point in that direction.

[100] Carvin, 1983, p. 292.
[101] See chapter eight 'Evident to the senses'.
[102] Schwobel in Webster (ed.), 2000, p. 33.
[103] Holwerda in Plantinga & Wolterstorff (eds.), 1983, p. 285.
[104] Weightman, 1995, p. 275.

CHAPTER FOUR

TORRANCE'S PROPOSAL – A NEW OBJECTIVITY

A new objectivity

In describing and accounting for the commonalities and the differences between general and theological science, Torrance's exposition centres upon conceptions of objectivity.[1] By construing older notions of objectivity as objectivism, he is able to formulate a new objectivity, centred on the notion of faithfulness to the object of the science, in which the object itself reveals the nature of its relationship to us (and thereby the methods of its respective science), rather than having our pre-established objectivity foisted upon it.

As Colyer points out[2] the scientific predicate of Torrance's theology should not imply any overarching *scientia universalis* or crude mimicry of any other *scientia speciales,* but rather in the fact that "theological science, like every science [*scientia generalis*], *has its own particular requirements and material procedures determined by the unique nature of its object or subject matter.*" This can be confusing for the more casual reader: often the title 'scientific theology' suggests a style of theology which fashions itself in analogy to a natural science, or one which makes significant play upon the results of some other specialist science. For example, for A. and O. Ritschl theology becomes a science in the eighteenth century by a process of absorption within historical science.[3] However this is precisely the sort of mimicry which Barth sees as, in fact, denying theology scientific status. So for Torrance the term means the opposite: that by being true to its distinctive object, theological science will often look *less*, at least superficially, like other specialist sciences. For Torrance it is the relation of subject and object which is the key to understanding the nature of science, natural or theological, and his narration of this relation's historical context is constituted by a series of oscillations in which the proper equilibrium

[1] For 'objective' being more or less a synonym for 'scientific' see opening lines of Scheffler, 1982, p. 1, and following, for a theologically impartial survey of the relationship between science, objectivity and subjectivity.

[2] Colyer, 2001a, p. 22, his italics.

[3] Pannenberg, 1976, p. 259.

of the relation has been unbalanced on alternating sides. When the subject has achieved dominance over the object – to the extent that a different object, a self-reflection or projection, becomes the focus (or at least occludes the proper focus) of theological study – then theology becomes non-theological.[4] When theology attempts to render God as an object like any other, without reference to his sovereignty, uniqueness or mystery, and pushes past analogical caution to fit God into a world of already known objects and concepts, then often despite appearances (for here theology is *too ready* to borrow conceptualities and methods from other sciences) it becomes un-scientific. We take each pole of the oscillation in turn.

* * *

Firstly, John MacMurray recalls his "incredulous amazement" at hearing "an English scholar describing science and religion as the expressions respectively of the intellectual and the emotional aspects of human consciousness." His amazement was against the background of his own and Torrance's Scottish Calvinism, in which "central stress fell upon sound doctrine."[5]

In Protestantism Barth traces this tendency to the Schleiermachian 'treaty'[6] between science and theology, which by rooting theology in our 'feelings of absolute dependency' institutes the 'heresy' integral to liberal Protestantism.

> The significance of Schleiermacher consists above all in his having, in his doctrine of Christian piety as the essence of the Church, given this [modernist Protestantism] heresy a formal basis which fulfilled the time preceding him, as well as foretold the time succeeding him. He is not the inaugurator, he is the great ripe classic of Modernism, which, if it understands itself, will never let itself be turned away from following him.[7]

However Barth was famously ambiguous about his relationship with Schleiermacher, whom he greatly respected,[8] but ultimately judged to have founded a tradition, to a certain extent without Schleiermacher's consent,[9] which led to the liberal Protestantism which Barth so

[4] McDonald, 2000, p. 15.
[5] MacMurray, 1995, p. 5.
[6] Barth, 1972, p. 410.
[7] Barth, CD I/1, p. 38.
[8] Torrance relates that the only time he saw Barth lose his temper at Basel, was when some students had rouged up Schleiermacher's bust. McGrath, 1999, p. 45.
[9] Barth, 1972, p. 410.

forcefully understood himself to be reacting against. The ambiguity hinges upon the relationship between object and method. For according to Barth, Schleiermacher was absolutely correctly dedicated to the idea that method is determined by object, so important at the founding of the University of Berlin (1810), but failed to see that therefore method must follow content rather than precede it.[10]

In beginning Christian theology with human feeling, however theological that feeling, Torrance too felt that Schleiermacher had made God 'mute', and that Feuerbach's charge, that such a theology was in fact anthropology, was justified.[11] No clear and proper distinction and connection had been made between the feeling of dependence, and the object upon which that dependence was based.[12]

For Torrance, Schleiermacher and his legacy marked one of the decisive forks in modern theology – one prong of which was ultimately to make theology one of the social sciences, dealing with religion as a human phenomenon – and a 'better' prong in which theology was an objective, positive, and scientific account of the reality of God's revelation in Jesus Christ and thereby of God himself. The centrality of Christology to Torrance's theological schema contrasts to the christological emphasis in Schleiermacher, which whilst fulsome, somehow seems to avoid arriving at a christocentric destination,[13] to the extent that Schleiermacher can say that the "Christian faith has no stake in the resurrection."[14] This Torrance traces to the Kantian influences on Schleiermacher, by which, according to McGrath, the prevenience of reason over revelation has made Christ simply an exemplar of an independently known system of ethics and knowledge of God, rather than as the source of these things.[15]

[10] Sonderegger, in McDowell and Higton eds. 2004, p. 194 – an idea preserved quite explicitly in Bultmann, 1984, p. 48.

[11] Torrance, 1969, p. 32.

[12] Whilst Gadamer posits the view that intuition and indeed interpretation had in Schleiermacher their roots in this 'feeling', Frei disputes that feeling has any place in his hermeneutics. Frei, 1974, p. 299.

[13] Torrance, 1990, p. 35.

[14] Frei, 1992, p. 75.

[15] "The idea that Jesus Christ should have any real impact on Christianity was strongly resisted by Enlightenment writers, whose commitment to the notion of a universal rationality and set of moral values led them to deny that Christ could have any constitutive role in shaping Christian thought. He might be permitted to illustrate or embody already existing moral values or religious ideas; he could not, however, be allowed to determine them. That privilege was to be accorded to the universalities of human reason alone. The idea that Christianity possessed distinctive ideas that were

If Torrance's complaint against Schleiermacher is that he has not *started* with God,[16] then Torrance is left with the problem, that alone amongst the sciences, theology has a starting point, the existence of which, by his own admission is neither self-evident nor indeed provable. If however his more correct assessment is that Schleiermacher, or at least those who followed him, have not proceeded scientifically, in that they have failed adequately to penetrate from phenomena to structure, he then can have no argument with his starting point – it is where he (or his followers) take it or fail to take it which is open to criticism. After all the physical sciences do *start* by considering phenomena which have only the merest connections with that which will *become* their 'objects' of study. On encountering the physical world, or the world of living organisms, the lay person, particularly the lay person from a non-Western culture as Feyerabend points out, will describe something with the merest commonalities to a text book of physics, chemistry or biology. The more advanced the book, the less it will contain 'everyday' terms, and deal with specialized concepts and terms unfamiliar to the layperson. According to Feyerabend the theoretical pluralist, this diversity of phenomenal description resides in incommensurable conceptual frameworks, frameworks which are largely human constructions. For the scientific realist, this is because mature science has been successful in penetrating below surface phenomena to elemental forms (Einstein's '*Raffiniert ist der Herr Gott, aber boschaft ist Er nicht*' – Torrance approvingly notes, set above the fireplace in the Mathematics Department lounge at Princeton), so that once so matured, it carries out its activities by referring to those forms, and subsequent discussion is directed by the nature of those terms, increasingly abandoning the initial phenomenal descriptions present at the outset of the investigation. The paradigm, as it progresses, will increasingly refine its concepts in a manner proportionate to its results – although the good critical realist will always see the possibility of a paradigm shift, as it were, out of the corner of his eye, and not rush to be over specific in his concepts.

For Polanyi through to Popper (i.e. almost anyone apart from Feyerabend) one sign of the success of paradigms is that their starting

true was dismissed as a self-contradictory statement. Christianity might echo or reflect the rational faith of the Enlightenment. But if it was *distinctive*, it was for that reason *wrong*." McGrath, 2002, p. 298.

[16] Or Barth's equally problematic assertion that we must "start with it and return to it" (CD I/1, p. 415) – here referring to Christ's deity.

points, even within discussions by the scientific laity of their host cultures, will increasingly feel at liberty to utilize the terminology of that paradigm – for Feyerabend, this proves the danger, for this is simply a propaganda victory by which science builds up a cultural heartland; for critical rationalists it is the due rewards of epistemic success and a culture wide 'field test'. So electrons, genes, the sub-conscious and DNA for instance, become themselves the objects of study, text books introduce them as starting points, and they become 'everyday' words, despite the fact that no one experiences them very directly and that physical science did without them for millennia.

Schleiermacher's programme could therefore be more kindly construed not as a betrayal, but as seeking a starting point, which enjoyed a degree of consensus – playing the Enlightenment game perhaps, but for good reason, and not ultimately in order to deny the centrality of God revealed in Jesus Christ to theological discourse. Torrance's task as a theological scientist, would then be to propose a theological paradigm, which by being maximally adequate to the phenomena offers theological science a maturity, so that its starting point could henceforward be Torrance's desired object: God. However the problem is that theological science has not achieved that maturity, because Schleiermacher's starting point, has not found a unique explanation, but a pluriform series of explanations – psychological, sociological, and theological. Each of these are therefore left to stamp their feet and talk past one another,[17] demonstrating again and again their adequacy to the phenomena, but ignoring equally valid explanations and failing to address the rather more interesting question of whether it might be possible to decide between them (and that criticism holds good for all parties). Pannenberg following Heiler is quite right to point out that religious claims both have a right to be taken seriously and that "a mere phenomenology, psychology of sociology of religions cannot get to grips with religion's specific object", but also that to be taken seriously it cannot be permitted to "accept[ing] such truth claims unexamined."[18]

Torrance is therefore at liberty to develop theological science strictly in the sense of the study of God, provided he allows that at a certain point a crucial assumption has been granted. It is highly doubtful that a 'crucial experiment' will be devised to discern the paradigm which

[17] A prelude to a Kuhnian crisis, Kuhn, 1970, p. 108.
[18] Pannenberg, 1976, p. 364.

best explains human religiosity. Here ironically, Torrance's criticism of Schleiermacher, is closer to the position of Feyerabend, than the scientific realist. Torrance is effectively saying that Schleiermacher's starting point has determined his destination.[19] However the scientific realist points out that the eventual 'object of study' transcends the choice of initial phenomena. What determines how each science unfolds is therefore not necessarily our initial phenomena or hunches, but as Torrance himself says, the fallibility and contingency with which they are held – if they are prematurely hardened into axioms, then the sceptical tension of fallibilism is slackened, but if they are held with an appropriate degree of 'fluidity', then even unlikely starting points can lead to scientifically realist and faithful concepts. Schleiermacher and the school that followed him have failed to be rigorous in the scientific task of pushing past inter-subjective phenomena to their underlying elemental order, but that does not entail an error in the choice of initial phenomena. Like Swinburne,[20] Torrance can argue that theism, even specifically Christian theism, *is* the most adequate to phenomena – but he cannot simply demand that Schleiermacher or anyone else need start their studies there. This does however admit of a rather mechanical account of scientific progress. We shall later on, when considering how natural science employs fluid rather than static axioms (chapter eight), see that whilst starting points do not logically make any difference, in the world of human research carried out with finite human resources, good initial 'hunches' will be a tacit personal component of competent scientific method. However in this particular case the high degree of consensus (that there are feelings of 'absolute dependency' that we might commonly call 'religious feelings') warrants Schleiermacher starting where he does.

There are of course, as Torrance the Barthian must admit,[21] fields of theological inquiry, particularly within Biblical studies, which can perfectly well consider various technical textual, sociological and historical questions with scant reference to or implications for the doctrine of God. Torrance however would resist any retreat to a ghetto, where *only* such questions can be asked, and where more substantive questions about the God referred to in those texts will be ignored for the sake of

[19] Torrance, 1969, p. viii.
[20] Swinburne, 1991, p. 287.
[21] Watson in Webster (ed.), 2000, p. 66.

academic respectability. As Soskice points out one of the reasons for the "popularity of the narrative theology paradigm is that it effectively brackets the question of truth."[22] Such a retreat is entirely arbitrary and decidedly unscientific, for it is perfectly proper for theology, being faithful to itself, to ask substantive questions about the being of God, and his relation to the world. Whilst such questions obviously open up the field to a great variety of religious traditions and revelations, Torrance is at no point concerned with apologetically defending the Christian revelation within that field, past demonstrating its coherence and rigour, and that variety cannot constitute any barrier to asking those questions, anymore than, for instance, the multifarious approaches and philosophical commitments found within literary criticism need imply that it is somehow improper to study literature critically. Though of course, the difference is that no literary critic doubts that there is such a reality as books.

Surprisingly perhaps, my conclusion at this point is therefore that Torrance has not here been realist enough. All too often Christian theology has fragmented into various schools, which beaver away without returning to the substantive questions which divide them. This does not however mean a return to the old fashioned foundationalism of establishing the existence of God, and his revelation in Jesus Christ, on neutral ground, prior to more substantive theological work. Barth and Torrance are right to dismiss that as a dead end. They are right to demand that it is only within a framework of iterative and spiralling critical realism where Christian theology can seek immanently to determine approaches which are epistemologically maximally efficacious.

However beginning with a phenomena such as human religiosity, or indeed others such as human suffering or hope, has the advantage of inter-subjectivity, without necessarily marking a betrayal at the outset, provided it is remembered that the scientific realist is always seeking to push beyond and behind the phenomena. It is the refusal to do the latter which marks unfaithfulness to object, for objects are known only as mediated by phenomena, and a refusal to seek beyond the phenomena would condemn theology to non-science, the essence of MacMurray's complaint.

---

[22] Quoted in Van Huyssteen, 1997, p. 189.

Correspondingly in the initial iterations of physical scientific process, the transition is made from the crude sensory description of phenomena, to scientific observations, in which phenomena are described in more theoretical but useable terms, progressively embedding observations within the emerging theoretical framework of that science. Feyerabend would see that as the process by which theory, particularly well established theory puts its stamp onto phenomena, even surprising and new phenomena, to bring them under its control.

* * *

Secondly whereas academic theology may identify itself in contradistinction to natural science (i.e. ignore the similarities between theological and natural science), so it may also over-identify itself with the natural sciences (ignore the differences between theological and natural science, such as God's personal nature).

One of the defining properties of enlightenment thought was a high regard for the progress of natural science, both for its great explanatory power and also with the technological marvels which followed in its wake. Thus, in particular, physics was seen as paradigmatic of good scientific method, and often when people casually refer to the methods or techniques of science, it is those of physics which they have in mind. To accrue some of this cultural capital for themselves, other disciplines began to imitate the methods and techniques of physics, and the extent to which they were able to do this, rested in good measure on the quantifiability of the object involved. Other features of the physical sciences, which were judged to have aided their success and therefore to be worthy is possible of imitation, included inductive inference from controlled and repeatable experiments; strict application of logic and mathematics in deductive inferences; univalent definition of technical terms; construction of neutral ideal observers and a distrust of any results based on tacit or personal knowledge; the creation of a geographically and culturally dispersed expert community which could rigorously peer-review published results; freedom from political, commercial, religious or other ideological factors within research institutions. It is easy to recognize here features which define our modern concept of the 'academic' and 'scientific', which have been examined by both Feyerabend and Torrance, in the latter's case with particular reference to the work of Polanyi. These features were sometimes rather crudely applied to other sciences with different objects of study, without due regard for the nature of that object. This is not just special pleading for theology, but also applies to the human sciences and even some physical sciences other than physics.

It is often impossible to construct controlled and repeatable experiments, for instance within the social sciences,[23] paleontology or volcanology – abduction rather than induction is vital in many sciences.[24] For theology, God's revelation, so insist Barth and Torrance, is not repeatable, controllable, or geographically or chronologically homogeneous – though the Biblical witness, once textual criticism has been worked through, provides a public constant – data – upon which systematic theology can work. Indeed that text with variants suitably annotated, forms an observational constant which Feyerabend is reluctant to grant to natural science.

Furthermore in many sciences the object and associated phenomena simply do not lend themselves to easy quantification. There have, of course, been movements within the human sciences, by those such as Kempski, attempting to assign mathematical procedures a more overarching function "from theoretical physics to linguistics."[25] However we must be wary of such approaches – ultimately the motivation will be convenience and a thirst for unambiguous answers, but it is likely that the search for truth may well be more long-winded. As Torrance points out, the logic of number however great its elegance and utility will here be misapplied, and the logic of word, organism, persons, grace or beauty will be the correct one[26] – indeed for Torrance it is vital that the development of a 'theo-logic' is a task performed *within* theology,[27] not prior to or detached from it. In chapter seven ('Religious and scientific language') we shall note the difficulties of precisely defining theological terms and concepts. The myth of the ideal impersonal observer will be debunked in the next section by the personalist philosophy of Torrance, Polanyi and others.

---

[23] Lacey in Archer et al. (eds.), 1998, p. 492 – Barth uses the analogy of scientists' abductive study of craters from previous shell impacts for theological concern with revelation – see McGrath, 2003, p. 162.

[24] Polkinghorne, 1998, p. 47; McGrath, 2003, Vol. 3, p. 153ff; Moore's assertion of the repeatability of experiments in natural science is too restricted to be anything close to universal – his 'disanalogy' on such grounds does not work.

[25] Pannenberg, 1976, p. 120.

[26] Torrance, 1969, p. 205ff.

[27] Its 'interior' or 'inner' logic, Torrance, 1969, p. 7.

Religious experience is by its nature, personal and private, and though others have made the case for its epistemological importance,[28] this case is not made by Torrance, primarily because of its radically subjective nature, and also no doubt his particular emphasis on the priority of the Biblical material in his Reformed context. Though obviously, religious experience is highly diverse and personal, it is illogical and unfounded to deny it any epistemic cash-value, when plainly for many individuals it is an important part of their noetic development. The work of for instance, Alston, is of interest here, but given that our focus is on Torrance for whom such an approach was alien, I shall not pursue it in any depth.

Feyerabend also identifies the technological marvels of Western medicine as superficially impressive, but not necessarily making that tradition any more successful in its results.[29] Furthermore it may rig the show with regard to what constitutes a marvel, and undervalue the significance of religious experience:

> Do you imagine a prophet or an early Christian or even an average Dogon being much impressed by two men stumbling around on a dried out stone [the moon] when he can talk to the creator himself?[30]

Whilst other criteria we listed – the creation of a diverse expert community and freedom from extraneous ideology for instance – may be reasonable aims for theological and other sciences, we have shown that scientific status cannot be achieved simply by aping physics or the other natural sciences – indeed such a crude approach may well undermine such an attempt. Torrance therefore carefully analyses the following similarities and differences.

Similarities and Differences

Torrance's most substantive exposition of the commonalities of general scientific method and scientific theological method, and the specifics of theological science, are found in *Theological Science,* and are summarised from p. 116 thus:

---

[28] Swinburne, 1991, p. 254; see also Griffith-Dickson, 2000, p. 133f; Alston, 1991.

[29] Feyerabend, SFS, p. 175, and also TDK, p. 89f.

[30] Feyerabend, TDK, p. 89; also AM, p. 307.

a) Theology displays "devotion to its proper object."
b) A scientific theology is what the Germans call *wissenschaftlich,* that is a rigorous, disciplined, methodical and organized knowledge.[31]
c) It searches for 'elemental form' from the variety and complexity of its data and results.
d) Its questions are fashioned and directed by what is already known about the object of study.
e) A 'problematic form of thinking', in which "our indicative statements are economically reduced to skeletal form."[32]

Then five which are specific to theology (p. 131ff):

i. The utter Lordship of the object which cannot be compromised.
ii. Respect for the personal nature of the object.
iii. A secondary objectivity in which He gives Himself to us in human form within our space and time.[33]
iv. Supremacy of Christology in our knowledge of God.[34 35]
v. Theology's demonstration of itself and its value can only be consonant with the unique nature of its object.[36]

The prominence of the conception of 'objectivity' is self-evident.

---

[31] Torrance, 1969, p. 116.
[32] Torrance, 1969, p. 126.
[33] Torrance, 1969, p. 136.
[34] Torrance, 1969, p. 137, see also Osborne, 2006, p. 368 on 'The Christological method'.
[35] "The Christian doctrine of God is to be understood from within the unique, definitive and final self-revelation of God in Jesus Christ … Every doctrine [of the Christian faith] will be expressed in its inner coherence with Christology at the centre, and in its correspondence with the objective reality of God's self-revelation in Jesus Christ." Torrance, 1996, p. 1.
[36] A contemporary formulation of scientific theology, comes from Alister McGrath. If Torrance has a uniquely influential predecessor in Barth, then he also has such a successor in McGrath, formerly Professor of Historical Theology at Oxford University, now at King's College, London University. McGrath is Torrance's biographer, and his writings are peppered with references to Torrance. The opening three volumes of McGrath's own systematic theology are entitled: *Scientific Theology,* and the shorter one volume 'consumer version': *The science of God.* The first volume of the former is dedicated to Torrance. In volume two (*Reality,* p. 236) he gives his own formulation of scientific theology:

(a) Theology displays "devotion to its proper object."

This first criterion is characterized by the epistemic virtues of obedience and faithfulness (theological resonances noted) to object (*kata physin*), the core of the general predicate 'scientific'. The proper object for theology is God, as he is revealed in Jesus Christ, not some lesser knowledge or lesser deity, but true knowledge of true God.[37] Neither is it knowledge of any human religious epiphenomena,[38] which must always be distinguished from actual knowledge of God, the former being ultimately knowledge of man. This theme constantly recurs in Torrance's work, and a quotation from the preface of another work, suffices as an example of its significance for him:

> The claim is put forward that theology is a pure science of a realist kind operating on its own proper ground and governed by its own proper object, and comparisons are drawn between theological science and natural science in these respects. Thus understood theology is a positive and progressive inquiry into the knowledge of God proceeding under the determination of his self revelation.[39]

Here Torrance is dependent on the achievement of Barth, which even his reluctant adversary Brunner says "without qualification ... or competition" is to have "given back to Protestant theology its proper theme and subject matter."[40]

So whilst Torrance and Barth concur that the proper object of theology, at least dogmatic theology, is God, Ricoeur argues that Bultmann,[41]

"Four major characteristics of a scientific theology [are that it] ... 1/ takes the form of a coherent response to an existing reality, 2/ is an *a posteriori* discipline, 3/ takes account of the unique character of its object, 4/ offers an explanation of reality."

McGrath then adds to these a fifth postulate, that scientific theology is Christocentric. McGrath is sticking quite close to Torrance's scheme, and it is difficult to stake out any clear water between them. Roughly 1/ corresponds to a-c, 2/ also covers a, 3/ corresponds to i-iii, and 5/ to iv. 4/ corresponds closest to b and d, though the notion of 'explanation' is uncommon in Torrance.

[37] Hence the concern noted by for instance Heron (*Homoousious with the Father*) in Torrance ed., 1981, that any accommodation with Arian Christology will have perilous consequences for our knowledge of God.

[38] Torrance, 1969, p. viii: "Theology ... in which we are directly engaged with knowledge of the reality of God and not just with religious phenomenality. Whenever religion is substituted in place of God, the fact that in religion we are concerned with the behaviour of religious people, sooner or later means the substitution of humanity in the place of religion." Also Pannenberg, 1976, p. 265 and Popper, 1969, p. 346.

[39] Torrance 1985, p. xiv; also Marley, 1992, p. 16.

[40] Brunner in Brunner & Barth 1946, p. 17 (opening words).

[41] To be fair Bultmann, 1984, p. 53, identifies exclusive concentration on *fides qua creditor* as the error of liberalism and the opposite concentration on *fides quae creditor* as the error of orthodoxy.

whilst agreeing on faithfulness to object as a guiding principle, counter-proposes that the object of exegetical work must be the confession of faith of the early church.[42] Pannenberg points out the unjustified antithetical disjunction here posited by Bultmann and Kähler.[43] Singular devotion to object does not entail devotion to a single object, as though objects existed in hermetically sealed boxes – Bultmann's, or perhaps more accurately liberalism's approach is analogous to a scientist, perpetually reliant upon inherited data, pedantically saying: 'It is the experience of x, y and z, that electrons have negative charge' rather than simply 'electrons have negative charge.' Certainly one is entitled to suspect and investigate rogue data, and the distorting effects of interpretation and transmission – the above formula *is* used in scientific research to signal such suspicions: 'Whilst the studies x and y lead their authors to suggest a link between cancer and …' – thus the liberal disjuncture leads the reader to suspect: 'Whilst the preaching of the primitive church was that God …' The question of whether that preaching contained true statements about God's nature, is one which must be taken head on, not side-stepped.

In science we investigate the mode of transmission, not to 'bracket the question of truth', but to clarify it – if we find the data to be suspect or to contain features inadmissible to 'modern man', then we must spell those out and draw clear conclusions. Torrance therefore understands such a distinction as a distortive focus upon anthropological collateral, obscuring the object of faith, whereas in fact the task of the scientific theologian in evaluating that collateral is to clear the lens of that faith, by factoring in aberrations in the data. The distinction is comparable to studying optics, without having any theories about the nature and properties of light, and focussing solely on the construction of lenses. Assessing the extent to which the Biblical witness as a record of faith, rather than simply as witness to revelation, distorts or occludes knowledge of God, is the task of Biblical theologians. It is found immanently within their hermeneutical task, and the danger of both conservatives and liberals is a presumptive prior explication.

Torrance's conception of scientific objectivity as faithfulness to object has been taken up subsequently by others such as Gadamer, but its chief antecedent for Torrance is Kierkegaard in his realignment of enlightenment ideas about the subject / object relation. Torrance

[42] Ricoeur, 1981, p. 59.
[43] Pannenberg, 1968, p. 22.

begins *Scientific Theology* with a substantial quotation from, and praise for, James Brown's 1955 book *Subject and Object in Modern Theology*, and it is his outlook which Torrance adopts. Brown notes that the antithetical approach to objective knowledge as truth and subjective knowing as mere opinion, is derived from, but moves beyond the teaching of Kant.[44] There then lies a fork in the road of modern theology with Kierkegaard, where one interpretation of his famous 'truth as subjectivity',[45] leads onwards through Schleiermacher towards Bultmann, ending at Cupitt,[46] and another (more authentic, according to Brown and Torrance) one,[47] which Torrance attributes to Barth, confronts and then avoids the simple Kantian antinomies of object and subject.[48] Such antinomies all too often play out in terms of divergent epistemic stances (realism and social constructivism), whereas Torrance insists that object and subject describe the two poles of a relation between oneself and the other:

> All our personal and rational activity operates with two poles, a subjective pole and an objective pole. The subjective pole is the knowing, believing, acting person, and the objective pole is 'the other', either another person or some reality independent of himself.[49]

'Objectivity' is for Torrance, the demand that whilst our being as subjects might properly seek to frame the object in terms conducive to the nature of our being, nonetheless the prevalent factor in our knowledge will be the nature of the object, which our semantics, conceptualities, ontology, epistemology, rationality and scientific method must all strive to conform themselves to, and not vice versa.

Torrance likens Kierkegaard to Einstein, identifying them both as thinkers working with fluid axioms.[50] In this identification with Einstein, a thoroughly realist and objective thinker, Torrance goes on to imply that Kierkegaard does not adopt a thorough-going subjectivity, as some interpreters suggest, but rather locates the subject, as the observer in relativity theory, within our logical and physical space.[51]

---

[44] Brown, 1955, p. 27 & p. 13.

[45] Kierkegaard, 1992, p. 131: "the difference is simply that science and scholarship want to teach that becoming objective is the way, whereas Christianity teaches that the way is to become subjective, that is, truly to become a subject."

[46] Using precisely that quote see Cupitt, 1980, p. 41.

[47] Brown, 1955, p. 194.

[48] Brown, 1955, p. 79.

[49] Torrance in Torrance (ed.), 1980, p. 11; see also Torrance, 1969, p. 311.

[50] Torrance, 1965, p. 73.

[51] Torrance, 1965, p. 73.

Barth,[52] Polanyi,[53] Scheffler[54], and Torrance[55] all line up, alongside the personalist tradition of Ebner, Rosenzweig, Buber, MacMurray[56] and Cullman, to insist that true knowledge is objective but personal, and that no contradiction or sliding into subjectivity is thereby implied, because it is utterly grounded in the reality of the other – it is to borrow a term of MacMurray's, 'heterocentric'.[57] Thus the way in which knowledge is understood as personal is firstly in the Kierkegaardian reinvigoration of the knowing human subject.[58]

Vitally, as Polanyi continually reminds us, there is nothing necessarily antithetical about scientific and personal knowledge. Michael Polanyi emphasises the personal as universal within our quest for knowledge,[59] and Torrance concurs:

> We are engaged in impartial, objective scientific thought, we are not eliminating the personal-coefficient, for that would eliminate science altogether; we are engaged in modes of thought which methodologically exclude unwarranted subjective features from influencing our observations and judgements and thus from intruding into the material content of our knowledge. This is what we mean by the 'disinterested' and sometimes 'impersonal' approach.[60]

'Personal' in this sense, should not be understood as 'private', as knowledge given privileged status and unavailable to the 'outsider' – knowledge that is largely *determined by the particular identity of the subject*. Any such distinction is fair neither to the natural scientific community nor to the Church. The discovery of the neutrino or the human genome are only in a very limited sense 'publicly available' let alone 'evident to the senses'. Rather they are researched, tested, developed, formulated and disseminated by a community of experts. Feyerabend alerts us to the possibility of dangers within such an (expensive) schema.

Similarly in Torrance's Barthian emphasis on *Church* dogmatics, we detect a positive and communal, if distinctively Reformed approach, in

---

[52] Webster in Webster (ed.), 2000, p. 9.
[53] Gunton in Torrance (ed.), 1980, p. 96.
[54] Scheffler, 1982, p. 2.
[55] Torrance, 1985, p. 14.
[56] In particular see his Gifford lectures: *The form of the personal.*
[57] MacMurray in Fergusson (ed.), 2007, p. 199.
[58] Palmer, 1969, p. 52, 178.
[59] Torrance, 1985, p. 132; Torrance (ed.), 1980, p. xv; Polanyi & Prosch, 1977, p. 195.
[60] Torrance, 1969, p. 303.

that the words of scripture, become *verbum Dei,* within the pneumatalogically inspired reflections of the ecclesial community, gathered to hear the Word addressed to them.[61] It is to that scientific community, and not to *particular* persons granted either authoritative ecclesial office or charismatic authority, that this work of theology is done, and then set forth to the world, not apologetically certainly, but at least as a witness to God's gracious self-revelation in Jesus Christ.[62]

The communal nature of the scientific enterprise is not however, to rule out a moment of personal breakthrough – a 'here I stand, I can do no other' moment – but that moment can equally well be found in the life of an Elitson, Galileo, or Harvey as a Martin Luther – a thought of Polanyi[63] with which Popper concurs.[64] Epistemologically, the non-Christian is on a par, according to Torrance, with the person who overhears one side of a telephone call.[65] He can learn something about the conversation. Indeed he might be able to surmise, with some degree of precision, its nature, but there would also be an element of mystery as to the full content of the conversation.[66] Moreover he is free, according to Torrance, at any point to take up the receiver and fully immerse himself, but essentially Torrance's theology is a 'theology of believers'.[67]

A radical critique of (a) comes from Feyerabend. Feyerabend's pragmatic philosophy married to his political liberalism implies that human understanding should pursue *above all else* freedom, and that 'truth', 'rationality', 'objectivity', are just the idols of false religions deflecting us from our primary purpose,[68] the theological resonances being in fact unhelpful imperial remnants:

> Reason, at last, joins all those other abstract monsters such as Obligation, Duty, Morality, Truth and their more concrete predecessors, the Gods, which were once used to intimidate man and restrict his free and happy development: it withers away.[69]

For the later Feyerabend these 'propaganda' terms have become demonised to such an extent that his only option is to deconstruct

[61] Rae, 2005, p. 130ff.
[62] Torrance, 1969, p. 210f; Osborne, 2006, p. 137.
[63] Polanyi, 1998, p. 308.
[64] Popper, 1992, p. 38.
[65] Torrance, 1969, p. 151.
[66] Phillips in Nielsen & Phillips, 2005, p. 46.
[67] Luoma, 2002, p. 132.
[68] See Preston, 1997, p. 192.
[69] Feyerabend AM (first edition), p. 180; see also Feyerabend, FTR, p. 11; see Cupitt, 1980 p. xi, for a similar metaphor for religious faith.

them, and empty them of meaning.[70] Like Rorty he assumes that the search for truth lies in basically the opposite direction from that of freedom: "substitute Freedom for Truth as the goal of thinking."[71] Feyerabend opposes the Enlightenment notion that properly understood truth would lead to freedom and freedom to truth, also a Biblical notion (John 8:32) to which Torrance would subscribe. So whilst Torrance is attempting to retain the tradition of 'objectivity' within science by constructing that objectivity in a certain way, Feyerabend wishes to deconstruct and dispense with such a notion altogether.

Feyerabend diagnoses the same initial condition as Torrance, though his prognosis is more serious:

> They extol the 'rationality' and 'objectivity' of science without realising that a procedure whose main aim it is to get rid of all human elements is *bound to lead* to inhuman actions.[72]

The italics are mine and demonstrate the necessity of the relation[73] – rigorous rationality and objectivity are not, according to Feyerabend, the solution to science's compromised autonomy, but part of the problem. Such words as reason, objectivity, and rationality, which invoke notions of scientific rigour, have become for Feyerabend simply words, propagandist terms which "connected with almost any idea or procedure … surround it with a halo of excellence."[74] However the necessity of the relation – humane science is good science – can therefore dangerously imply its reflexivity – good science must therefore be humane science, undermining Feyerabend's view that science must be judged ethically, regardless of epistemology.

Feyerabend is most in error in characterising the search for objectivity and rationality as simply 'getting rid of personal elements', for the proper alignment of object and subject has humanitarian significance in its epistemic prioritising of 'the other' over against our self-concerns. The relation of subject and object is not precisely mapped to the relation of human to inhuman – the main danger of adopting an outlook "that pleases *us* for some reason or another", is that this may well lead to very inhuman actions towards *other* people. Such a construal of the relation diminishes the realist possibility of the other curtailing our

---

[70] Feyerabend, FTR, p. 10.
[71] Rorty, *Contingency, irony and solidarity*, p. xviii.
[72] Feyerabend, FTR, p. 299.
[73] The danger Feyerabend is talking about, in terms of moral objectivity, Barth foresees – 1933, p. 479.
[74] Feyerabend, FTR, p. 10.

freedom and choice, a curtailment which offers the other a proper integrity and freedom of its own, independent of our desire. Objectivity is not the diminution of the human, but the abasement of the subject, so that the object, divine, human or otherwise might be known in the depth of their reality.

Torrance's less radical prescription (understand oneself in relation to the reality of the other) is more subtle and useful than Feyerabend's (make decisions by whether 'they please us for one reason or another'). We can sum up the position thus:

- Both Torrance and Feyerabend diagnose the enlightenment confusion of objectivity and objectivism.
- For Feyerabend this leads to a false and blunt dichotomy between objectivity (i.e. objectivism) and relativism. A dichotomy between 'cantankerous twins' which has infected western thought from Plato.[75]
- Feyerabend's solution: plump for 'relativism' as a 'sticking plaster', meanwhile demolishing the 'straw man' of objectivism.
- Torrance's solution: reconstruct a more subtle objectivity. Only oppose objectivity to relativism when they are crudely understood.

Torrance is therefore more subtle in showing that the ideal of 'objectivity', whilst easily demolished if constructed as a straw man, is capable of reform, development and rehabilitation.

(b) A scientific theology is what the Germans call *wissenschaftlich*, that is a rigorous, disciplined, methodical and organized knowledge.[76]

The danger of this criteria is that it is liable to become like voting for mother's love and apple pie – after all few are happy to have their science described as lacking in rigour, undisciplined, unmethodical, and disorganised. Luckily for our purposes one of those few is Feyerabend, who sees most of these as propaganda criteria, hijacked and put to use by Western natural science. If human beings are properly to flourish, it may be the case that scientific knowledge cannot be rigorously and

[75] Preston in Preston, Munévar, Lamb (eds.), 2000, p. 93; see also Bernstein, 1983.
[76] Torrance, 1969, p. 116.

methodically tidied up and organised, particularly if it has to deal with multiple efficacious but incommensurable paradigms and traditions. Perhaps language is the best analogy – we may well have an entirely legitimate *desire* to see our language 'tidied up', but only in very artificial contexts can that actually be achieved without curtailing that language's activities and abilities (as Wittgenstein saw), so that generally we shall have to live with language as it is, warts and all. Any attempt to force language into tidying itself up is doomed to result in only an artificial and oppressive Procrustean conformity. Feyerabend's assessment of the value of such criteria is descriptive – there *has been* no single thing identifiable as the 'scientific method'[77] or 'rationality',[78] so we cannot make non-anarchic prescriptions – what we must do is simply limit the humanly undesirable consequences of moral or rational absolutism.

However for Torrance the order of our knowledge is not an order *imposed* by man as Kantians maintain, and indeed in that our knowledge is critical and partial, we may have to content ourselves with a degree of uncertainty, partiality and disorder until clearer relations are revealed, but the unitary order inherent and discoverable within layered ontic strata is underwritten by the unity of the act of creation. It is therefore both non-contingent and also rests upon a particular theological doctrine. We do not impose our organisation, rather science can only progress 'under grace', whether we acknowledge it or not, with the assumption that an intelligible natural order awaits our discovery of it, having been previously breathed into life by a creative God. That assumption underlies and motivates the scientific quest itself – a quest with a strong doxological component.

Thus the principle, that it is the object of the sciences which must immanently dictate their methods (Torrance), is distinct both from saying that we choose those methods (Feyerabend),[79] or that there exists a universal method or rationality. There are at least three options here, which Feyerabend tries to shut down to two. If we are agreed that there is no self-evident universal method, then the question still remains whether the methods we end up using, are freely chosen, forced upon us by human authority, or inherently discovered within our investigation. Whilst for Feyerabend a corollary of this diversity is

[77] Paul, 1982, p. 16.
[78] Feyerabend, AM, p. 18 & Preston, 1997, p. 170.
[79] Feyerabend, RRSM, p. 19.

to make a virtue of inconsistency,[80] the genuine epistemological capital of it is gained by Barth,[81] but not at the expense of anarchy.

(c) It searches for 'elemental form' from the variety and complexity of its data and results.

This methodological principle Torrance identifies with Ockham's razor.[82] Ockham's razor is concerned with the principle of simplicity, *simplex sigillum veri,* that is, for instance, such an important principle in the work of Swinburne.[83]

In this section it is the work of Einstein which Torrance[84] singles out as a model of simplicity, in the beauty and economy of the 1915 field and geodesic equations of general relativity theory. Therefore one should not mistake this understanding of simplicity as ontological and conceptual economy, with simplicity as easily-understood – relativity theory is not generally regarded as the latter. Simplicity may also be cheaply achieved by ceasing rigorously to correspond our concepts faithfully with reality. Though Dirac famously valued 'beauty' over 'experimental fit'[85], Torrance would be wary of this short-cut to simplicity. However on the other hand, Torrance is also keen the opposing short-cut which equates the search for elemental forms with a too simplistic reductionism, unwarranted by the nature of the object.

Feyerabend sees in this principle the danger that existing theoretic structures, claiming to have objectively found the things being searched for, then impose themselves upon the variety of phenomena, natural interpretations (see 'Theory and Observation' in chapter five for a description of natural interpretations) allowing us unconsciously to ride rough-shod over that diversity, and so he would reject this conservative demand in favour of theoretical pluralism.

[80] Feyerabend, SFS, p. 191.

[81] Barth, CD I/1, p. 8.

[82] Torrance, 1969, p. 117:*'pluralitas non est ponenda sine necessitate'* or *'frustra fit per plura quod potest fieri per pauciora'* though as Torrance notes in a footnote, and Boehner in his introduction (Ockham, 1957), the usual form *entia sunt multicplicanda sine necessitate* does not actually occur in Ockham.

[83] Swinburne, 1991, p. 55–6. However Swinburne makes a connection between the role of simplicity in explanation and simplicity as an attribute of God. Whether this connection can be properly made is debatable, and this is not the claim that I or Torrance are making – here simplicity is solely related to explanation, and is not referred to in the sense of a divine attribute.

[84] 1969, p. 118.

[85] E.g. Scott, 1995, p. 12.

Feyerabend and Torrance are here essentially talking at cross-purposes. Torrance's theologically motivated scientific realism leads him to *assume* that ultimately the search is for a uniquely economic description which is adequate to the phenomena, whereas Feyerabend is simply not interested in refining one paradigm down to its most economic form, but proliferating preferably incommensurable paradigms, firstly to be adequate to the phenomena, but also to avoid any particular paradigm suppressing society's cultural variety.

Most philosophers of science are forced to admit criteria other than empirical adequacy in theory assessment, generally scope and simplicity, if only to avoid rampant *ad hoc*-ness. However it follows that if these are non-reducible, then they are liable to function as *a priori* philosophical commitments – after all why should nature be economic rather than profligate with her 'forms' (excuse the anthropomorphism), when at the phenomenological level clearly she is not? The answer of the critical realist is that simplicity is not an end in itself, to be pursued without constraint, but that the scientist merely seeks to describe reality as simply as that reality will allow inherently of itself.

Furthermore, there are no quantifiable universal measures of simplicity (Ockham assumes one is choosing between two or three quite similar alternatives) or scope, and even where agreement is reached, it is unlikely that one theory will win awards in all categories, in which case there is the further problem of criteria ranking.[86]

Feyerabend would rather we abandon notions of criteria – why 'conquer abundance' in this way? For him simplicity is simply a political choice over and against pluralism, and not a self-evidently scientific one. Again we note the emergence of a pattern:

- The rationalist (narrow sense) assumes things (here order and simplicity) which cannot be rationally demonstrated.
- Torrance: In such cases we must extend our notion of rationality, including theological commitments, if reality demands it, even if we cannot construct a water-tight *logical* warrant. This is not irrational but extra-rational.
- Feyerabend: We must abandon rationalism.

[86] We have already seen that this is an issue in modern physics: relativity and quantum theory score empirical successes, but they lose out in compatibility.

(d) Its questions are fashioned and directed by what is already known about the object of study.

In our encounter with God's revelation we find that ultimately (like Job, and often by Jesus in the Gospels, e.g. Mark 11.30) every question is answered by a question directed at us,[87] for ultimately, as Barth puts it, God "is unadulterated answer."[88] Neither can we simply devise our questions beforehand and ask them in the order of our choosing, for each answer, must force us to reformulate our questions – 'kinetic thinking' as Torrance labels it.[89] Therefore what might properly *function as* foundational and axiomatic within our system, must for the sake of scientific rigour be held up to scrutiny within this encounter – 'fluid axioms' are required, so that no prior system dominates the order or content of the questions we pose, or how we interpret the answers.

In natural science, an equivalent faithfulness to object means that ultimately everything must be revisable, down to our very conception of time and space.[90] In pure science the shape of future research must always be contingent upon emerging results.

The human or theological scientist must furthermore be alert to where the nature of the object is personal (see differences (i)), for here, our questioning may not be met with passivity, but with a more active response, that forces us to reconsider our next move. In particular with God, our response must also be doxological. Upon questioning God, we find that our primary bearing towards him is that of worshipper rather than observer.[91] Part of such adoration (Einstein calls it 'awe'[92]) is a sacrificial bearing, even a *proper* partial *sacrificium intellectus*,[93] towards our conceptual categories,[94] this being as much part of a properly scientific disposition within the natural as the theological sciences.[95]

---

[87] Torrance, 1969, p. 26.

[88] Barth, CD II/1, p. 154; "Here we have an objectivity profounder than any objectivity natural science has to do with, for we are face to face with the creator of the universe itself, encountering us as himself in Word and Act, who cannot be mastered by us, but who remains in his transcendent Reality the Lord over all our questions, before whom we are at the bar, questioned down to the very roots of our being" – Torrance, 2001, p. 154.

[89] Torrance, 1965, p. 73.

[90] Torrance, 1965, p. 73.

[91] Alston, 1991, p. 187.

[92] Torrance, 1969, p. 122 & 184.

[93] Bultmann, 1984, p. 52.

[94] "To sacrifice themselves, over and over again, to the community's narrative texts, to the 'new language'" – Surin, *Turnings*, p. 217, cited in Patterson, 1999, p. 65.

[95] Pannenberg, 1968, p. 185.

Scientific objectivity is therefore related to the theological idea of kenosis:

> He has no means of measurement of *what it is* that God's self-revelation is, other than God's self revelation itself. The fact that there is no such means of measurement other than the object of measurement itself is testimony to the Godness of God. Man must set himself aside and rely utterly on God revealing himself: 'He must increase, while I must decrease'.[96]

This understanding of all science as posterior comes on the side of theological science from Barth, but equally from the side of natural science from Michael Polanyi, who according to Feyerabend,[97] "is one of the few twentieth-century scientist-philosophers to notice and assert the research-immanence of scientific standards."[98] So Feyerabend,

---

[96] McDonald, 2000, p. 347.

[97] Feyerabend, KSR, p. 218.

[98] Polanyi, 1998, p. 170; Polanyi certainly is an upholder and critical friend of the scientific establishment (Polanyi, 1998, p. 216; Polanyi, 1998, p. 220; Polanyi, 1969, p. 56–7 & p. 66), and uses the metaphor of farmers tilling different strips of land to underline the enlightened self-interest which should underpin the rigorousness of this system of control (Polanyi, 1998, p. 217). However whereas Feyerabend construes this as a failure to carry through insightful discoveries to uncomfortable conclusions ("This is indeed the conclusion that has been drawn by Michael Polanyi, one of the few twentieth-century scientist-philosophers to notice and assert the research-immanence of scientific standards. According to Polanyi there is no way in which outsiders can judge science. *Science knows best*." KSR, p. 218), I would submit that this is a prime example of Feyerabend's tendency to demolish straw men: by showing the impossibility of producing universal *a prioris* with regard to the scientific method, Feyerabend believes that this entails the iconoclasm of *Against Method*, whereas in fact a more measured Polanyian approach is simply to assert that such methods are immanent rather than prior, and then get back to work. Feyerabend offers no convincing argument against this.

We need to bear in mind how Feyerabend's approach can be and was portrayed by his opponents: as advocating political interference with scientific independence. Feyerabend gleefully admits that he is guilty as charged, whereas Polanyi (Polanyi, 1969, p. 65) argues for upholding that independence. A difficult calculus is involved, beyond the scope of this book: scientific independence vs scientific accountability – but I would venture that the smart money is on the former, and one must understand the contexts within which Feyerabend and Polanyi worked – it is one thing for a democratic and responsible government to ensure that taxes are well used and minority viewpoints represented, it is another for a totalitarian state to coerce science in the oppression of its own people. On one side there are the armament and pharmaceutical corporations who are so powerful and rich that it is a constant struggle for governments to impose proper regulation, and on the other Lysenko and Galileo – the question is never scientists or governments, but what is good science and what is good government.

So whilst Preston, commenting on Feyerabend's methodological pluralism says that, "there is a strong connection here with the work of Michael Polanyi," (Preston,

whilst conceding that Polanyi (and Kuhn, also Putnam[99]) has noticed this immanence, criticises him for failing to draw his anarchic or indeed democratic conclusions, therefore ultimately collaborating in natural science's imperialist demands for pre-eminence and money.[100] Feyerabend points out that not only are scientific questions posed within a certain structure, they are also posed with certain ends. The pure scientist, particularly the realist pure scientist is aiming at objective truth. Feyerabend contends that far from being a noble aim, it would be more useful if science could content itself with the more modest aim of promoting human flourishing. Pure science is to become practical science – only in this way can it be ethical science. Again the pattern is recurrent: for Torrance the truth will enable human flourishing, for Feyerabend it is simply a rival thrown up by a particular and rather bossy tradition.

Feyerabend concurs with Torrance's desire for freedom from prior system, but not because the object of our investigations demands to be

---

1997, p. 172) Lakatos, Feyerabend's rationalist sparring partner, defends Polanyi's moderately conservative conclusions, against the radical conclusions of Feyerabend (Lakatos, 1978, p. 176). So too Torrance is in agreement with Polanyi and Feyerabend that "there is no one scientific method which is universally applicable" (Torrance, 1969, p. 106, also McGrath. 2002, Vol. 2, p. 283), but sides with Polanyi against Feyerabend in what conclusions that entails for scientific practice.

Torrance agrees with Feyerabend's methodological pluralism, but denies that the posterior method must inevitably collapse into methodological anarchy.

Polanyi (Polanyi, 1998, p. 60ff) argues that as scientific research proceeds, so certain results occupy either our focal or subsidiary awareness – at any particular point, we have a 'question in hand' which we regard as focal, and a slew of supporting theories and results which we hold as subsidiary – our focus may change at any point, in response to new developments or unexpected results. Not only do we elect a 'question in hand', but the objectivity of the other is always liable to throw up unexpected results, which scientifically *demand* that we investigate further and account for them. At such a point, though we might defer them, we cannot ignore such questions indefinitely without scientific cowardice. Polanyi says that it is rather like poking tentatively at something with a stick (Polanyi, 1998, p. 61), and Torrance likens it to using a probe (Torrance, 1992, p. 5).

This approach leads Torrance to his most concise definition of: "scientific theology is active engagement in that cognitive relation to God in obedience to the demands of His reality and self-giving" (Torrance, 1969, p. ix). Active engagement in this sense is to be understood as a consuming and passionate (*not* dispassionate) regard for the object, and not in the sense of *intellectus agens,* or as Torrance prefers *intellectus possibilis*, which when left unbridled, caused so much mischief within both Roman and Neo-Protestant theology, corporately in the former, and individually in the latter, becoming the errant and dominant self-assertion of man's knowing (Torrance, 1965, p. 69,).

[99] Tambiah, 1990, p. 143.

[100] Feyerabend, SFS, p. 99; AM, p. 295.

heard and considered, rather because the human questioner has a right to make whatever enquiries he wishes. Recurrence again of an old theme: once each has shaken themselves free of the enlightenment search for a prior and universal rationality and method, they then both turn to rather jealous gods: Torrance to the reality of the object of inquiry, Feyerabend to human freedom and flourishing. Often the commands of those gods will coincide, but by no means always. Moreover, once any *a priori* method is ruled out, Feyerabend gives scant consideration to how an immanent method might yet offer a viable alternative to anarchy.

(e) A 'problematic form of thinking', in which "our indicative statements are economically reduced to skeletal form."

Here Torrance uses some of the material from (c) and (d) to introduce to us the problematic, as opposed to interrogative, mode of inquiry, in which we are "not concerned with new knowledge, so much as with clarifying what we already know"[101] – though we may not know *how* it is known, and as we attempt to put it into orderly form, the nature of the object may mean that the simple logical processes and forms of mathematics (if a, b, c, … then x) may be denied us. Therefore though our aim will be a "clear account of the steps taken in discovery and required in verification,"[102] the nature of the object may limit our ability to complete that ambition. Furthermore though we may seek and often be able retrospectively to spell things out in this way, it will "rarely represent the actual way in which convictions arise."[103]

Again Torrance's thinking is closely related to the outlook of Gadamer: "There is no such thing as a method of learning to ask questions … a question occurs to us that breaks through into the open and thereby makes an answer possible."[104] Gadamer then elides this idea, with another, which is a favourite motif of Feyerabend: inauthentic dialogue, says Gadamer "engages in dialogue only to prove oneself right and not to gain insight,"[105] likewise Schmidt points out "in an authentic dialogue each is open to what the other has to say."[106] However

[101] Torrance, 1969, p. 130.
[102] Torrance, 1969, p. 126.
[103] Torrance, 1969, p. 127.
[104] Gadamer, 1975, p. 365.
[105] Gadamer, 1975, p. 363.
[106] Schmidt, 2006, p. 112, 159 – dialogue as opposed to Ricoeur's theory of text.

Feyerabend is not interested in finding or pursuing such open dialogues, but simply in castigating natural science for its pre-emptive closure of dialogue, and then assuming that traditions are incommensurably stuck within their self-referentiality.

What this form of thinking amounts to according to Torrance, is '*the logic of Grace*',[107] a dialogical rather than dialectical method, in which revelation in theological science plays a role analogous to discovery in natural science.

At this point we open up an important distinction between the thought of Feyerabend and Torrance with regard to rationality and logic. For Feyerabend, to act 'rationally', is to act according to some universal and predetermined set of logical rules (we can clearly see that Feyerabend is erecting a straw man here). For Torrance, to act rationally is to temper one's logic (like selecting a tool) to the nature of the object, so that rationality will have a multi-layered aspect, and the particular mode of rationality or logic will be immanent within the science of that object. For Feyerabend such a tottering construction cannot stand, it must come down. Torrance would probably agree – were it not for God, by his covenantal creation gifting the necessary rigidity to the structure.

Torrance's stratified conception of rationality(ies) and logic(s) is given stability by external theological underpinning (see quote 111), whereas for Feyerabend no such remedial remedy is possible once the enlightenment form – singular universal – of these is dismantled. If one removes the theological support from Torrance (pre-empting an important conclusion of this book), one may well open the floodgates to Feyerabendian anarchy – all one is left with *is* a plurality of rationali*ties*/logic*s* – and Torrance's schema means he rules out any alternative non-theological escape route. Feyerabend's swipe at universal rationalists – "once Feyerabend justified his last-minute lecture cancellation with the remark that a female critical rationalist argued him into bed"[108] – fails immediately the more sophisticated rationalist points out that seduction has a different rationality from diary keeping. Smedes points out that the logic may be different, even when the language is the same,[109] so care must be taken to operate with logical connections as determined by the object and not simply by the text

[107] Torrance, 1969, p. 128.
[108] Hoyningen-Huene in Preston, Munevar & Lamb (eds.), 2000, p. 7.
[109] Smedes, 2004, p. 176f, because you are 'playing a different game'.

referring to the object. As Barth reminds us "Christian faith is not irrational, anti-rational, not supra-rational, but rational in the proper sense."[110] For Torrance:

> It is because all rationality in the universe has a single source in the creator, and because therefore the different modes of created rationality, number, word, organism, and beauty all interpenetrate one another, that there is a profound unity in all the arts and sciences, but in the nature of the case a unity that will appear properly only in stratified structures of openness to the unlimited reality and uncreated rationality of God.[111]

However for Feyerabend (see chapter five, 'Against Method and Farewell to Reason') the first clause is undecided at best, and given that Torrance refuses to prove or justify God's existence in any conventional sense, we are left simply with 'appearances' – the appearance of 'stratified structures' – which lack any underlying unity, stability or reference in a Creator God. Within Feyerabend's anarchic epistemology we shall see that he moves from a rejection of universal method to a 'farewell to reason' – rationality becoming simply a tribal deity invoked by the tribe of rationalists. At this point all that distinguishes Torrance and Feyerabend flows from granting or not granting Torrance his theological premiss – or fideistic assertion if one is being hostile.

Feyerabend fails here to mount an attack with bite, because the force of his assault is so focussed upon the straw-man of a rather naive *a priori* enlightenment universal rationality, that the bullets whistle over the head of more recent critical rationalists, who insist that method and rationality do not precede content, neither are they raised sovereign above it, but are discovered within it and mediated by its host tradition. Feyerabend's stated intention is surprisingly similar to Torrance's: "Combining the two elements [naturalism and idealism] we arrive at the idea of *a guide who is part of the activity guided and is changed by it*."[112] However Feyerabend is then too concerned to ward off positivistic and Popperian 'idealism', to explicate how the path between idealism and naturalism might be negotiated and to give examples of it. Feyerabend may have pre-empted MacIntyre's popularisation of the idea that rationality and justice are inescapably tradition-mediated entities,[113] and that a rational debate is a particular

---

110 Barth, 1949, p. 23.
111 Torrance, 1985, p. 24.
112 Feyerabend, AM, p. 232, his italics.
113 Palmer, 1969, p. 183.

form of guided encounter between two traditions and need not be the only or the best one. However if, despite initial differences of appearance Achtemeier succeeds in showing a deeper convergence between Torrance's critical realism[114] and MacIntyre's tradition mediated rationalities,[115] then it appears that Torrance may have succeeded in squaring Feyerabend's circle. Rather than building prudently on his admittedly incisive capital, Feyerabend is too often enticed to invest in rather wild and untenable rhetoric. Patterson's distinction between a critical realism in which rationality is solely the possession of the intellect (perhaps as a workman owns the tools) and a post-critical realism (which she attributes to Torrance) in which such rationality is at least partly imbedded in the object, may frustrate Achtemeier's attempt at harmonisation, but as both realisms at the epistemic level agree that the nature of the object will render some modes of rationality more efficacious than others, it is here unlikely that such a distinction will decisively affect our conclusions.[116]

(i) The utter Lordship of the object which cannot be compromised.

Torrance continually underlines that the unique nature of the object of theological science means that what we gradually uncover in our investigations is our own creaturely status under that object.[117] Furthermore it is here, above all, where the importance of the proper exercise of scientific fallibalism and scepticism rightly understood, incarnated in the theological virtue of humility, is paramount, so that the unique authority of the divine subject will in no way be compromised or trespassed upon.

God's lordship implies, in epistemological terms, that basic theological beliefs are foundational for all knowledge, not in the sense necessarily that they determine that knowledge, but in that they prescribe the freedom under which that knowledge is attained. "Christ's universal Lordship should determine our account of the relationship between philosophy and theology."[118]

Human reason, however powerful a tool, must still be understood as gift and servant, and not as weapon of self-assertion by which we

[114] Colyer, 2001a, p. 325.
[115] Achtemeier, 1994, p. 371; see further p. 208.
[116] Patterson, 1999, p. 13.
[117] Torrance, 1995, p. 346.
[118] Moore, 2003, p. 123 – thus Moore also seems to fall into 'type four'.

"escape from self-inflicted immaturity."[119] Scientific objectivity cannot therefore consist of our attempt to circumnavigate our position as finite creatures within creation. Torrance detects within classical natural science an imaginative platonic leap towards the impersonal which attempts to transcend this relation,[120] a tendency which Kirby traces to Descartes.[121] Thus we have seen how Torrance attempts, with mixed success, to show how discoveries in modern physics (relativity, quantum theory) and mathematics (Gödel's theorem) have led to a realignment of that relation.

However as we saw in connection with relativity theory, this is to mistake the imaginative component of the idealizations of the Cartesian opposition of subject and object for ontological ones – components such as the mathematical diminution of a solid body to infinitesimal point, a mathematical fiction acknowledged and known, or the posing of an imaginative series of 'what ifs' – and I see no reason to judge such exercises as illegitimate within science, so long as they are understood as useful techniques and not ontic descriptions. Crucially their appropriateness within each science will be dictated by the on-going investigations into its object, so that if techniques harden to a universal objectivism (as Feyerabend and Torrance argue) then this betrays objectivity, but if it remains as a tool with specified parameters of applicability, then its use is valid. Frictionless surfaces and non-elastic collisions never occur in the real world, but the mathematical models drawn from them are important tools for describing that world, as long as such descriptions are not taken therefore to ascribe actuality to such events. Such imaginative diminutions are not simply an Archimedean attempt at the impossible, but attempts to give priority to the reality of the object. All 'objective' sciences are governed by the procedure of reducing the "unwarranted subjective features" towards their zero limit – this does not entail the actual realization of that possibility, obtaining that absolute singularity where the 'personal-coefficient'[122] is obliterated, which is the mistake of objectivism.

Thus whilst the Lordship of the object of theological science is established almost by definition, it must, more than Torrance allows, be penetrated right down to the ground of its mediacy, one step away

[119] Kant, *What is Enlightenment?*, opening words, p. 140.
[120] Torrance, 1985, p. 14.
[121] Kirby, 1992, p. 125 – see (ii).
[122] Torrance, 1969, p. 303.

from mystery. For even here scientific procedure demands that we are free to enter into the 'what ifs' of revelations apart from the Christian one, though we withhold any act of commitment. Imaginatively to enter into the paradigms of other faith communities, is possible, just as it is mathematically possible to envisage various solutions to Einstein's equations (known as metrics). Such fallibilism needs to be held in tension with faith commitments – credal anathemas are generally unhelpful[123] in the scientific pursuit of theology, whatever ecclesiological and political purposes they might serve.

For Feyerabend too, as for Torrance, such imaginative exercises are liable to become acts of self-fulfilment, whereby, perhaps unconsciously, patterns of thought are woven deeper into noetic frameworks than we intend: hence impersonal results lead to 'inhuman actions.'[124] The danger – and here Torrance and Feyerabend are both right, though they do not express it as well as they might – is that as theological and natural scientists imagine themselves into another paradigm or into a thought-experiment, they are liable to delude themselves into believing that they have transcended their prejudices and preconceptions and arrived at an impartial meta-level of judgement. However a physicist trying to understand how voodoo works (Feyerabend), a Christian trying to understand what a Buddhist believes, are still physicists and Christians doing useful thought experiments – the danger lies in thinking of themselves as having transcended such identities and staking out the objective neutral territory as theirs. Feyerabend has a point therefore when he say that we cannot simply impose our standards of scientific objectivity of others, for those very standards are bred from within a particular set of procedures and methods. However as ever with Feyerabend there is a difference between setting out the pitfalls in what might quite possibly be a rather noble attempt (to assess traditions objectively), and insisting instead that we abandon any such attempts, and satisfy ourselves with rather unwieldy and sometimes unpleasant alternatives.

Feyerabend's complaint that "hardly any religion has ever presented itself just as something worth trying,"[125] is unfair whilst he maintains his insistence on the 'immersion' technique of the anthropological

[123] Moore, 2003, p. 148.

[124] Feyerabend, FTR, p. 299.

[125] Feyerabend, AM, p. 218 – for contrary view see Phillips in Nielsen & Phillips, 2005, p. 5.

method. Firstly, the claim is contestable, and secondly, it is inconsistent with one of his few methodological principles. Thought experiments, theological and scientific, are possible and desirable and need not threaten either Torrance's realism or Feyerabend's humanitarianism, in fact they are essential tools in attaining these goals. God's lordship is known by our willingness to submit all of our knowledge – even our knowledge of him, for our status as creatures means that we are in fact in no position to compromise his Lordship – to imaginative testing.

(ii) Respect for the personal nature of the object.

For Torrance theology is a dialogical (rather than a dialectical) relationship between persons, and cannot be understood in any other way.

> It is personal relation which best describes the structure of reality itself, therefore the model of the personal agent must be primary in our attempt to think intelligibly of God, rather than the impersonal model of detached observer over against the object, with its unbridgeable gulf between subject and object, which has returned to exercise so much influence on Western thought since Descartes.[126]

As well as the relation being personal on the side of the observer, within theology as opposed to the physical sciences, knowledge is personal in a further, more important, sense: that it is *of* a person (the person of God).[127] Torrance rejects the Cartesian antinomy of object and subject, and adopts the personalist I-Thou and I-It categories of Martin Buber.[128] Our knowledge of God is our knowledge of his act of revelation, and whilst God cannot be known by anything other than revelation by and of himself – what Barth and Torrance call, following Athanasius and Calvin,[129] *Deus loquentis persona* – that is not to imply that revelation is a means of knowing unique to God. On the contrary, revelation is the normal mode of knowledge between persons, and thus we verify the personal nature of God.

> All knowledge of persons is by revelation. My knowledge of you depends not merely on what I do, but upon what you do; and if you refuse to reveal yourself to me, I cannot know you, however much I may wish to

[126] Kirby, 1992, p. 125.
[127] Kirby, 1992, p. 125.
[128] See discussion McGrath, 1999, p. 202–3, also e.g. Torrance 1969, p. 309.
[129] Barth, CD I/1, p. 304; Calvin, 1960/1, I.vii.4; McGrath, 1999, p. 144; See also Frei, 1974, p. 23.

> do so … for scientific method is based on the assumption that things are what they appear to be.[130]

The difference between knowledge of God and of other persons is one of power. Humans can be coerced into revelations – sometimes very subtly, as when carefully framed questions suggest desired answers; sometimes aggressively or violently – but God cannot be coerced. We are not even in the position of asking the questions, God in the ultimacy of his sovereign freedom, chooses, by his good grace and bound only by his loving purposes and nature, to let himself be known to humanity, and thereby becomes an object of our knowing, with no compromise to that sovereignty or freedom.

Therefore a certain subjective restraint is a necessary scientific precaution within the study of persons. It is all too easy to project upon other persons wishes and feelings which in fact lie within ourselves, and such projective phenomena extends to non-human animals, inanimate objects (by their divinisation) and the divine. I recently wrote a letter to a church newspaper to defend the title of a book written by a friend of mine: *If you meet George Herbert on the road … kill him.* It was in response to another correspondent who had found such a statement about a much-loved Anglican divine offensive. However the point of the title and the letter (if the correspondent had read as far as page four he would have found the explanation) is that it originally referred to the Buddha, and was written by a great Zen teacher – Lin Chi. The point Lin Chi was trying to make is that if you think you have met the Buddha, then be very careful that you have not been deceived, for it is far more likely that you have met your own projection of what the Buddha should be.[131] The divinisation of our own hopes, fears, needs, and those of the tribe, are a constant feature of human religiosity, but also occur within natural science. In singular devotion to the object of God, Christian theology must operate with severe suppression of such anthropological intrusions (Torrance's 'unwarranted subjective features').

Torrance leaves us here with an unresolved tension. For how is scientific Christian theology to extricate itself from the doxological privileges which its object properly demands? This tension presses on right

[130] McMurray, 1996, p. 89.

[131] J. Lewis-Anthony, *If you see George Herbert on the road … kill him*, Continuum, London, 2009.

to the question of the existence of God – for if this question cannot be considered without recourse to the facticity of God's revelation to us, what feature of that revelation is ultimately to distinguish it from self-revelation? Obvious self-projections – the superiority of our tribe, the desire for fecundity – are relatively simple to spot, but psychology tells us that the human mind has a more or less unlimited capacity for self-projection – how can scientific rigour circumnavigate that in the search for knowledge of an object which is extra-sensible? By far the most dangerously distortive projections, at least in the long term, are not those concerned with naked aggression, but those disguising more subtle and complex repressions.

Conversely, in Feyerabend the privileges of the personal subject are in danger of trampling wholesale over those of the personal object. For whilst the pleasure criterion, at worst, is liable to produce, in the long-term at least, an understanding of the external physical world which is dangerously self-delusory, it is far more dangerous, in that it is likely to produce short-term attitudes seriously detrimental towards other persons, based on one's own wishes and desires: hence the muddle we noted over anti-semitism. Scientific fallibilistic restraint over the voluntary and conventionalist components of our knowing is not only likely to yield a more accurate picture of physical reality, but more importantly help us to identify the personal nature of *other* objects and be open to *their* rights, needs and desires – and furthermore secure more accurate scientific information concerning them.

The anthropological method tells us that we cannot genuinely understand other persons without a degree of self-giving and immersion, but what it is liable to omit to tell us, especially if employed in reaction to a hard objectivism, is that a posture of imaginative self-forgetfulness is also a prerequisite for knowledge of other persons. Sympathetic resonances may be false, and a stiffer sounding board may allow us properly to discern fundamental frequencies. Other persons may be quite different from us, and the aim of the human sciences is to give those differences voice.

Scientific knowledge of persons is therefore founded, as we shall encounter Ricoeur saying of text, in holding in tension rigour and obedience, listening and suspecting. Passionate commitment[132] held

[132] Lest one should be afraid of admitting passions within science, we note Scheffler's 'cognitive emotions' and 'rational passions' – 1982, p. 139ff.

without prejudice is the tensioned stance we must take up. Feyerabend neglects the second pole, whereas Torrance is apt to render the personal as something rather un-passionate.

(iii) A secondary objectivity in which He gives Himself to us in human form within our space and time.[133]

Moreover according to Torrance, in our encounter with God's revelation, the derivative nature of our own subjectivity is exposed by the aseitic subjectivity of God,[134] who gifts us our subjectivity in his graceful act of creation.[135] God, the person whom theology wishes to inquire about, reveals himself, not only as the source of our creaturely subjectivity, but as himself indissoluble subject.[136]

It is true that God may *become for us* object, but that is a movement of condescending grace, not something which we can require, force or in any sense demand or guess of God, neither does it compromise, but rather underlines, that aseity.

The discovery of the derivative nature of our personhood is, according to Torrance, a guard not only against objectivism, but also against a radical subjectivism. Torrance believes that neo-Protestantism, in emphasizing the formulae 'Christ died for me', has unscientifically tipped the balance in favour of the 'for me', so that man in his self-knowledge and self-fascination, becomes the controlling pole of the relationship, whereas in fact the opposite is true: humanity is only truly known as found in Christ.[137] Tillich might be right in saying that so far as our systems and categories are called into question, so the question is 'man himself,'[138] but not in the sense that man or his religiosity is made the object of inquiry, rather so that if necessary they may be deracinated for their unfitness.

According to Torrance God *gives himself to us in grace as object*, but that is the only means by which he is (self-) objectified and his indissoluble subjectivity kenotically self-transcended. In that we are creatures of time and space, God meets us on this plane in Jesus Christ, so

[133] Torrance, 1969, p. 136.
[134] Patterson, 1999, p. 26.
[135] Torrance, 1982, p. 139.
[136] Torrance, 1994, p. 131.
[137] Torrance, 1969, p. 81f.
[138] Torrance quotes this himself from Tillich's *Systematic Theology II*, p. 14.

that by his assuming of our human nature, and not through any effort of our own, we might truly know God on his plane, the plane of eternal and real knowledge. Our cognitive knowing of God is thus itself utterly shot through with grace. Sin is not simply a matter of moral imperfectness, from which rational intellectual activity is immune or even a defence, it is also a cognitive defect transcended only by grace.[139]

Justification has an epistemological as well as an ethical and soteriological reference – epistemologically it insists that the only legitimate demonstration of Christian truth is that which is in accordance with its nature – i.e. by grace – and that to seek justification on any other ground, is not only fundamentally false in itself but is to "falsify the Gospel at its very basis."[140]

In contrast to Kant's (and Feyerabend's[141]) demand for maturity in order to know, for Torrance it is childlikeness and repentance, *even with regard to epistemology*, which is the key to true knowledge.[142] For both Torrance and Feyerabend there "is an ineradicable moral dimension to all human knowing,"[143] but this dimension plays out in opposite directions – for Feyerabend it is the responsible maturity of adults for our own knowledge, for Torrance it is the humble acceptance of our infantile dependency, even in the cognitive realm.

According to Torrance, God's freedom and sovereignty certainly entails that he *can* speak to us in whatever manner he wishes, but what is foundational for the Christian is that he *has* revealed and objectified himself for us in Jesus Christ, and the incarnation is thus the controlling centre of all scientific knowledge of God. The scientific realist is concerned with the actual rather than the possible or speculative.[144] As Pannenberg admits "as Christians we know God only as he has been revealed in and through Jesus Christ."[145]

God's self-objectification is thus properly a matter of history, and indeed therefore for historical investigation, though this should not be taken as warrant for the crude application of positivistic historical

[139] Barth, CD II/1, p. 70.
[140] Torrance, 1965, p. 163.
[141] Feyerabend, RRSM, p. 19.
[142] Torrance, 1969, p. 253, a result employed to existential ends in Bultmann, 1984, p. 122.
[143] Colyer, 2001a, p. 343.
[144] Langford, 1972, p. 159.
[145] Pannenberg, 1968, p. 19.

science to the incarnation. On the other hand, there can be no premature jumping of Lessing's ditch, leaving sensible history behind and living solely on the pastures of intellectual reason. This Torrance labels as '*hybris*'.[146] There can be no docetic embarrassment at this physicality:

> The event of revelation as described for us in Scripture has everywhere a natural, bodily, outward and visible component, from the creation, by way of the concrete existence of the people of Israel in Palestine, the birth of Jesus Christ, His physical miracles, His suffering and death under Pontius Pilate, His physical resurrection, right down to His coming again and the resurrection of the body.[147]

In condescending to be known by us, God in his grace,[148] accommodates his being into our knowing, indeed into our language[149] and finitude, not by their reaching out to grasp him, a task for which their poverty is all too obvious.[150]

Whilst Feyerabend develops some solicitude for theology, he also rightly points out how past, and indeed present, soteriological exclusivism so often leads to 'inhuman actions'.[151] The problem is that God's grace and accommodation demand on our behalf repentance and humility, qualities which while translated by Torrance into the scientific categories of objectivity and fallibilism, are for Feyerabend suspect, because the question will always be actually: *whose repentance and humiliation are required*? In other words can we translate the evangelical imperative into terms which are not characterised by the conforming of the standards of one tradition to the paradigms of another – the dividing line between honest open yet robust dialogue and the 'closed encounter' of insensitive evangelisation is an important one, which Feyerabend tells us concerns those meta-level standards all too easily imposed by 'objective' traditions of religion *and* science – in few other sciences than theology will faithfulness to object, and indeed personal integrity, have to be so carefully balanced with respect towards others operating within different paradigms.

---

146 Torrance, 1969, p. 137.
147 Barth, CD II/1, p. 265.
148 Barth, CD II/1, p. 70.
149 Torrance, 1995, p. 399.
150 Barth, CD I/2, p. 115.
151 Feyerabend, FTR, p. 299.

(iv) Supremacy of Christology in our knowledge of God.[152]

This point follows from (iii) for Torrance: God has decisively and epistemologically normatively revealed himself in Jesus Christ (John 14:7,9, also Mt 11:27, Lk 10:22, Jn 6:46, 7:29, 8:55, 10:15, 17:25[153]).

As Barth says, "God is not known and is not knowable except in Jesus Christ,"[154] so Torrance concurs:

> every doctrine [of the Christian faith] will be expressed in its inner coherence with Christology at the centre, and in its correspondence with the objective reality of God's self-revelation in Jesus Christ.[155]

To ask to 'see the Father' remains therefore an impiety (John 14:8–9), one which will have eschatological fulfilment, and whilst our own capacity for 'seeing' will be transformed so that we will be able to receive the fullness of God, no different or contrary information about God's nature will be imparted. We shall not then see something *other* than what is known in Jesus Christ, we will simply be fully assumed into the unity of Father, Son and Spirit. Scientific theology, for Torrance, is therefore simply unconcerned with the question of knowledge of God outside of Jesus Christ – as a science it is posterior to an actual encounter with an actual event, rather that dealing in the possible and speculative, and Athanasius' dictum is fundamental to the project: "It is more godly and accurate to signify God from the Son and call him Father, than to name him from his works and call him Unoriginate."[156]

Nowhere else does Feyerabend's charge that scientific ideas are bound to be incarnated in political realities, have more force for the scientific theologian than in adopting this highly christo-centric approach. The church's attitude towards Jews, heretics, dissenters, and people of other faiths, has been by turns violent, condescending, and exclusivist. Such a scar on the face of Christ's church needs little elucidation. However the danger is that Torrance's desire to render theology 'scientific' means that his accounts of all such encounters tend to be so blandly ideational, and stripped of their political and polemical

---

152 Torrance, 1969, p. 137; see also Osborne, 2006, p. 368 on 'The Christological method'.

153 List of references from Khang, 1992.

154 Barth, CD II/2, p. 509.

155 Torrance, 1996, p. 1.

156 Quoted several times in Colyer, 2001a, being on p. 374 the closing words.

dimensions, that they fail to answer arguably the church's most burning questions. More significantly scientifically, it leaves theology wide open to Feyerabend's charge that at each stage, often retrospectively, the orthodox party has effectively 'rigged the show'. If the Christian church is the scientific theologian's laboratory, it is a rather grubby one, and a primary task must be to show how it can be cleared up.

(v) Theology's demonstration of itself and its value can only be consonant with the unique nature of its object.

Our demonstration (equivalent to, but different from, verification in natural science) of the divine object therefore must itself be divine – only God can reveal God – non-theological proofs of his existence are bound to fail (see p. 239). However insofar as the incarnation takes place in time and space, so knowledge of God is revealed through a person in history, and therefore is known, at least partially, as historical, and by historical verification. To this end the Gospels are historical witness. Torrance notes that the word 'partially' does not imply a division of human truth (Jesus of history) and divine Truth (Christ of faith), an approach which Torrance simply regards as doubly errant,[157] but a knowledge of divine-human truth, where available to all humanity, bound together hypostatically in the one person of Jesus Christ, we are able to verify true man and true God. As Moore puts it, though obviously with different ecclesial priorities, "God's verification of himself is eschatological. It has taken place proleptically in Christ's resurrection; its consummation is anticipated in the eucharist."[158] Divine being and historical facticity are thus inseparably united in the ground of our knowledge and are not to be divided in our continuing inquiry into the fact of Jesus Christ.[159]

Feyerabend distrusts any notion of 'demonstration' or 'verification' on supposed supra-theoretical grounds, preferring pragmatic demonstration of value to any attempt at objectivity. Thus the 'universal intent' of Christian systematic theology, even if humbled by reduced circumstances, is an unwelcome feature.

What is more serious for Feyerabend is that Torrance's relegation of the ethical and historical as secondary to the ideational, therefore omits

[157] Torrance, 1969, p. 325.
[158] Moore, 2003, p. 122; also Pannenberg, 1976, p. 343f.
[159] Torrance, 1969, p. 325.

any 'demonstration' assessed against the criterion of human flourishing. In wishing, for ecclesiological reasons, to draw a fairly sharp and respectful distinction between the divine incarnation and the actions of Jesus' all too frail followers, Torrance is in danger of rendering *any* consideration of the 'fruits of the Spirit' redundant. The danger is that the Jesus Christ event is thereby robbed of any potency. Surely the veracity of the Christian gospel will and should be at least partly measured by the ability of Jesus' followers to live up to his ideals, enabled by his Spirit so to do. Such a notion, at least in St Paul, seems entirely compatible with his demand that justification is by grace.[160]

Conclusion

As we have seen Feyerabend binds with chains of necessity the notion of objectivity (understood as the 'removal of human factors'), with 'inhuman actions'. This is in reaction to modernity's tendency to equate scientific procedure, whether in the natural sciences, or indeed in historical science,[161] with the diminution of all elements of knowledge which are personal, preferring to render knowledge as something impersonal and abstract. Objective knowledge here is essentially viewed as being 'attached' to the known rather than the knower, and whilst this version of objectivity grants due priority to the object, its reflexive repulsion from the subjective assumes that all subjects, whatever their context and inheritance, will absorb that knowledge identically. However no such ideal Archimedean observer is possible, and even if we were able to specify objective knowledge by supra-linguistic means, it is certain that its absorption into the noetic medium of subjects with highly variegated world-views and contexts would necessarily incorporate distortive effects in any subsequent transmission of such knowledge. Torrance therefore rightly concurs that knowledge must be rendered appropriately to its knowing subject and in that sense must be known as personal. However conversely Torrance reminds Feyerabend that knowledge is primarily of something or someone, even if it is always by someone, and knowledge must therefore be faithful to the nature of that object. This objectivity, as faithfulness to object, is still opposed to the objectivism which treats every object as merely

[160] Patterson, 1999, p. 158.
[161] Brown, 1955, p. 41.

an object, assigning its place in the world according to some abstract and universal schema, heedless of the nature of the knower, and regarding the knowing subject simply as an epistemological nuisance. Thus according to Torrance, subjectivity and objectivity refer not to two different renderings of knowledge, in which the desire of the subject and the hard facticity of the object are constantly in conflict, but refer instead to the two poles of the act of knowing, subject and object, held in proper relation. This is the approach which he takes from Brown and Kierkegaard.

Feyerabend however, having asserted that the old ideal of objective knowledge is somehow bound to lead to inhuman actions, then too crudely appends human choices as the panacea for this epistemological ill. Whilst agreeing that the knowing subject should never be removed from the plane of knowledge, we cannot admit this drastic therapy, and instead decisively prefer Torrance at this juncture. For what is really liable to lead to 'inhuman actions' is not simply to be equated with the removal of human factors – what "pleases us for some reason or another" being a supremely human factor. What leads to inhuman actions (and we do not deny that the old scientific objectivity can be used to dress up our desires in neutral colours) is precisely the preferring of the subject's desires and wishes, over the facts of the case, particularly when the facts of the case include the desires and wishes of persons who are the object of this knowledge. Evil and oppression remain anthropological and indeed theological concerns, and therefore it is doubtful whether a programme of radical subjectivisation of scientific theory would make the world a better place, indeed it is highly likely that our search for knowledge would be severely hampered.

I have noted the importance which Torrance derives from the person-to-person nature of theology and the impossibility of isolating theology from religious ideology and context. The creation of a rigorous and catholic expert community within theology is a perfectly practical aim, which Torrance shares, and whatever their shortcomings, his 'distillations' of history[162] indicate that this need not simply be a contemporaneous community, but (as in natural science) a progressive tradition which is constantly sifting and accumulating results from the past. There is however also a disanalogy here. Whilst it might be too

[162] Hardy in Ford (ed.), 2005, p. 173.

simplistic to characterize natural science, even in its normal stages, as purely progressive, and theology as simply archaeological, it is nonetheless the case that, if, as Torrance says, revelation is analogous to discovery, then natural science is far more concerned with pushing forward to future discovery, and theological science more concerned with elucidating past 'discovery' (the problematic as opposed to the interrogative, to use Torrance's terms, or the abductive rather than the inductive, though as we have noted theology is not alone in being an abductive science).

The fundamental question to Feyerabend's critical challenge is then whether, by attacking foundationalism, he can succeed in showing that the necessary corollaries follow, in order to tip the balance between these similarities and differences, decisively towards the latter. If the content of *scientia generalis* can be shown by the following challenge to be too vague, insubstantive or incoherent, and the differences too marked and too much like special pleading for a plethora of *scientia speciales*, then the coherency of any overarching notion of the 'scientific' will be deconstructed and Feyerabend's challenge proved successful.

CHAPTER FIVE

FEYERABEND'S CHALLENGE – 'KNOWLEDGE WITHOUT FOUNDATIONS'

> One seems to be forced to choose between the picture of an elephant standing on a tortoise (what supports the tortoise?) and the picture of a great Hegelian serpent of knowledge with its tail in its mouth (where does it begin?). Neither will do.[1]

At the outset we noted the danger of presenting too progressive or programmatic an exposition of a self-confessed anarchist's thoughts, at the same time noting that unless there was a certain degree of coherence within a core set of ideas, those ideas become too slippery and inchoate to handle at all, and lack any relevance for academic debate. This anarchy (Feyerabend is keen to distinguish his epistemological anarchy from political anarchy – "not the most attractive philosophy"[2] – politically Feyerabend's anarchy entails libertarianism and democratic freedom) is the only medicine[3] strong enough to challenge the hegemony of Western science, but it is medicinal and therefore transient, and at a deeper level there is a coherency and continuity to Feyerabend's ideas. It is deployed to therapeutic rather than edifying ends[4]; it is a dose of salts rather than doctor's tonic.

With that rider, we here set out the core ideas of epistemological anarchy. They are anchored in the context of a discussion of foundationalism because they start with a critique of any absolute dichotomy of theory and observation – a feature strongly associated with the nadir of scientific foundationalism: logical positivism – and arrive at a voluntarist and relativist outlook which is the foundationalist's mirror image. The question will then be whether the starting point leads inevitably to that destination.

---

[1] W. Sellars, *Science, Perception and reality*, 1962, p. 170.
[2] Feyerabend, AM, p. 9.
[3] Feyerabend, AM, p. 17.
[4] Rorty's taxonomy – see Griffith-Dickson, 1995, p. 327; or Hacking's 'critical' vs 'speculative'.

Theory and observation: counter-inductivism, theoretical pluralism and the rejection of empirical cumulativism

Nearly all accounts of natural science are predicated upon the assumption that scientific activity consists of theoretical reflection, which explicates, orders and accounts for some sort of observational data recording natural or social phenomena. Logical positivism wished to give a systematic and foundational basis to such activity by isolating statements and/or terms which had observational content (basic / protocol sentences) from theoretical statements. The positivists then took a nominalist line with theoretical statements, denying that they necessarily aimed to describe anything with ontological reality, ultimately being only ordered descriptions of observational data. Science was relieved of what most had hitherto considered to be its naturally fundamental, if burdensome, task of economical ontological description, admitting instead that all such description was ontologically derivative, contingent, revisable and indeed disposable. Furthermore the demarcation of meaning, not simply of science, was restricted to such basic sentences, and theoretical statements which could be shown to have consequences directly verifiable by such sentences – any statements which failed this infamous verification criterion were judged not simply un-scientific but meaningless. The downfall of logical positivism is well documented, but we begin our exposition of Feyerabend's later work with his rejection of this division and therefore of the positivist schema.

Popper had relaxed the demarcation criteria, firstly by making falsifiability rather than verifiability the criterion of demarcation, and furthermore by drawing the lines of demarcation between science and non-science rather than meaning and non-meaning. However this too had been undermined by those such as Lakatos[5] and Kuhn, who had shown that theories, *even sometimes at their conception*, lived with anomalies, which only *became* falsifiers upon the advent of a new more empirically adequate theory. In times of revolution, competing theories might take considerable time to complete the contest, because assessing empirical adequacy was not always straightforward, particularly as other criteria such as scope and simplicity (lack of '*ad hoc*-ness') might also point in divergent directions. However any notion

[5] Lakatos, 1978, p. 3.

that observations straightforwardly determined theory was undermined by a number of factors. Firstly, according to the Duhem-Quine hypothesis, given that individual theories only exist within paradigm-scope theoretical complexes, it is never possible to isolate a single theory for testing, for there is always a possibility that an adjustment elsewhere in the complex might rescue the theory.[6] Observation is therefore 'theory-laden'[7] and "there is no such thing as uninterpreted fact."[8] Secondly and consequently, established theory – perhaps from completely unrelated fields – is incorporated into experimental equipment, assumptions, design and interpretation.[9] Thirdly there is the mathematical triviality that any finite set of data has an infinity of possible mathematical solutions.[10] These factors led to the wide spread realisation (including in fields other than natural science) that all observation is theory-laden or theory infected.

Feyerabend's own particular derivation of this principle is from his consideration not simply of natural science, but also of the linguistic theories of the later Wittgenstein, which suggest that meaning must be known holophrastically, never isolated from linguistic context (the language game). There can be no exemption for scientific languages, however artificially and precisely tailored to their application, because to be communicable they must be capable of correlation with natural language.[11] We cannot simply take against natural language as Frege wanted us to, because whatever its limitations, as Wittgenstein points out, human knowing is immersed within it.[12] The linguistic turn in philosophy, challenges the philosophy of science by suggesting that philosophy cannot help *frame* 'correct' definitions, but must confine itself to *understanding* language in all its variety.[13] The well known Messianic secret of the gospels may be understood as an example of an occasion where a context must be (re-)built *before* it is possible to

[6] Duhem, 1954, p. 187; see also Popper, 1969, p. 112.

[7] An important term for Feyerabend initially coined by N. R. Hanson in Hacking, 1975, p. 120.

[8] Gelwick, 1977, p. 80.

[9] Feyerabend, POE, p. 138.

[10] Popper, 1992, p. 138; Polanyi, 1946, p. 9.

[11] Pannenberg, 1976, p. 181.

[12] Feyerabend agrees with Wittgenstein (Wittgenstein, 2001, p. 22 & Wittgenstein, in a letter to Russell, and from the *Tractatus* in Anscombe, 1996, p. 161–3) that sentences do not simply give expression to a supra-linguistic meaning; Hacking, 1975, p. 178.

[13] Hacking, 1975, p. 7.

affirm a particular predicate of Jesus of Nazareth – not to mention other examples where predicates ascribed to Jesus, will be held with ambiguity, because the point is not that Jesus fits the category, but rather that he marks it out.

Feyerabend's outlook at the time of *Against Method* is somewhere (not very explicit – but as we shall conclude Feyerabend is hedging his bets here) between Wittgenstein's instrumental theory of language[14] and Sapir-Whorf's conceptual determinism, naming his thesis the "contextual theory of meaning."[15] I place Feyerabend *beyond* Wittgenstein's instrumental theory,[16] though Feyerabend claims he is simply *adopting* the theory, because I agree with Preston's assertion that Feyerabend "over-extends the theory."[17] As D. Z. Phillips has pointed out, the crucial question is whether language games and forms of life are capable of external criticism, reform and correction, albeit only by way of a mutually attentive dialogue, or whether they are fundamentally isolated and cut off from one another – this is also the question Feyerabend does his best to avoid. However Phillips is insistent against his critics, that Wittgenstein, himself, and indeed common sense demand that they are, even if the tenor of Wittgensteinian philosophy will often be one of pointing out the difficulties involved in so doing.[18]

Wittgenstein's theory, anticipated in Frege,[19] has been widely used amongst others by Waismann[20] and Quine.[21] Theologically both Ward and MacDonald have used it in their assessments of Barth, with obvious implications for Torrance. Ward maintains that whilst both Wittgenstein and Barth were non-foundationalists (at least in the enlightenment sense of the term), neither were they linguistic

[14] Wittgenstein, 1967, Sect. 43.
[15] Preston, 1997, p. 26.
[16] Floyd, 2005.
[17] Preston, 1997, p. 29.
[18] Phillips, 1986, p. 10 – the row between Phillips and Nielsen et al. over 'fideism' is perhaps best simply understood as an opposition of theory and practice – in theory, language games / forms of life are externally assessable but in practice they are nearly always utilised to show how a certain tradition cannot be assessed by another – so that they appear isolated. Perhaps if a worked out example was given by a Wittgensteinian of how Azande beliefs could properly be assessed from the point of analytical philosophy, then that would help.
[19] Frege, 1997, p. 90.
[20] Waismann, 1965, p. 313.
[21] Quine, 1960, p. 51.

idealists,[22] which is the ground Feyerabend is moving towards, and therefore the reason why I think Feyerabend is not true to Wittgenstein. Thus Ward and McDonald also implicitly concur with Preston's assessment that Feyerabend is pushing Wittgenstein's theory beyond its proper boundaries.

In 1962, when Feyerabend introduces the theory, his super-realism makes his original use of the theory true to that of Wittgenstein, but when he cuts the ropes with realism, his untethered destination is more akin to that of Sapir-Whorf. Whorf's theory (at least in its strong version) says that language is theory determined. The 'strong' version of the Sapir-Whorf hypothesis (though, perhaps sensing danger, Feyerabend avoids explicitly endorsing this version) promotes a strong conceptual relativism in which our linguistic structures, rather than reality itself, determines the 'facts' of our world:

> The fact is very significant for modern science, for it means that no individual is free to describe nature with absolute impartiality but is constrained to certain modes of interpretation even while he thinks himself most free .... We are thus introduced to a new principle of relativity, which holds that all observers are not led by the same physical evidence to the same picture of the universe, unless their linguistic backgrounds are similar, or can in some way be calibrated.[23]

The force and one-sidedness of this theory is one reason why Whorf in undiluted form has fallen out of favour within linguistics,[24] as well as being strongly challenged at the time.

That observation is theory-laden has now become a common place in the philosophy of science.[25] However whilst scientists have in one way or another found strategies to account for this fact, Feyerabend makes the stronger case, veering from Wittgenstein towards strong Sapir-Whorf, for observation being theory-*determined*:

---

[22] Ward in Webster (ed.), 2000, p. 281; though this was the error which dogged the *Tractatus* – Hacking, 1975, p. 84.

[23] Whorf, 1956, *On science and linguistics,* p. 214, It is worth noting that Whorf himself was drawn to this view partly by "the fundamentalist shadings of his Methodist Episcopal religious background" (Carroll in introduction to Whorf, 1956, p. 7) – Whorf however classifies all the Biblical and European languages under the Indo-European umbrella so that generally speaking in theology, the latter condition will not come into play.

[24] J. Jospeh & E. Koerner, *From Whitney to Chomsky: essays in the history of American linguistics*, John Benjamins, 2002.

[25] Van Fraassen, 1980, p. 14.

> Thesis I: the interpretation of an observation language is determined by the theories which we use to explain what we observe, and it changes as soon as those theories change.[26]

However Feyerabend never secures the case for this stronger and rather unlikely assertion – unlikely not least, because it would render paradigm shifts impossible. Perhaps such shifts might occur – and this is Feyerabend's point – purely on a whim or because we 'fancy a change'. However this is historically difficult to maintain, for whilst it may be an uphill struggle (analogous to the rise of an opposition party in a totalitarian state) new paradigms do on occasion definitively solve defects within hegemonous paradigms. Certainly natural science needs to hold in constant tension a fallibilism, which is open to new and surprising results, with a certain conservative caution which does not demand new theoretical paradigms with every anomalous result – anomalies can either be persistent (the perihelion of Mercury), but more often they are explained and absorbed by existing theory without an undue degree of *ad hocness*. Feyerabend is (again a standard tactic) playing on a weakness of logical positivism in order to defeat it – for he offers no precise account of what a scientific observation or a scientific 'fact' consists of – a problem which was to dog logical positivism itself.

Logical positivism requires a simple relation between 'observation statements / protocol sentences'[27] and sense data. At first observations were simply taken as consisting of 'sense data' – however it quickly became clear that scientists do not record sense data as such, and so logical positivism, firstly Neurath, then Carnap and Popper, were forced to attempt to construct a bridge, which was itself free of theory. Most, including Feyerabend, judge that attempt to have been a failure, and gradually most logical positivists returned to psychologism even if Popper continued to hold out. Even if the bar is raised from sense data to inter-subjective agreement, it is unlikely that such inter-subjectival groups will transcend divisions of culture and traditions, as few scientific peer groups are representative of the totality of human demographics, having been schooled and trained in the common

---

[26] Feyerabend, RRSM, p. 31.

[27] In an attempt to get around this problem Schlick muddies the waters by trying to distinguish protocol from observation sentences, making the former more basic descriptions of sense data (see Scheffler, 1982, p. 107f) – however the problem is a more general one, in that whatever terminology we use, the gap between 'raw feels' and scientifically useful observation statements is too large to be trivial.

and distinctive tradition required to set up the experiment and record the results.

When diversity is present, for instance in revolutionary situations, Feyerabend is explicit that such a scientific consensus, even over bare 'facts', is unlikely to exist. Returning to his primary historical example, Feyerabend notes that on April 26th 1610 Galileo was found in a foul mood by Horky, because between the twenty four professors gathered at Magini's house in Bologna to use the telescope, there was violent disagreement over what was *actually seen* celestially. The reasons were a complex compounding of psychological and technical imperfections.[28] "Science knows no bare facts ... they are essentially ideational."[29]

Whilst Quine continued to reduce the scientist ultimately to a processor of sensory input and output,[30] he concedes that "no experiment may be expected to settle an ontological issue; but this is only because such issues are connected with surface irritations in such multifarious ways, through such a maze of intervening theory."[31] It is precisely this maze which makes the task of logical positivism so untenable.

This is therefore a rejection of the 'bucket theory' of scientific observation,[32] in which like fishing with a bucket, everything is brought up for examination. Rather as in fishing using a net, whose size of mesh will determine exactly what is caught, so in natural science, we, prior to experimentation, nominate and isolate a particular part of reality which we wish to investigate, and by isolation and appropriate control procedures seek to determine the effect of particular variables: "Observation does not simply mean 'seeing' but rather 'seeing as.'"[33] Furthermore as observation becomes written data, the processes of initial interpretation and assimilation within existing theory are already taking place, whereas the logical positivist would like to seal off all this experimental process from contamination by theory. As Duhem admits:

> What the physicist states as the result of an experiment is not the recital of observed facts, but the interpretation and transposing of these facts

[28] Feyerabend, AM, p. 88.
[29] Feyerabend, AM, p. 19.
[30] Quine, 1990, p. 1 (opening words).
[31] Quine, 1960, p. 276 (closing words).
[32] Ritchie, 1948, p. 83.
[33] Smedes, 2004, p. 94.

> into the ideal, abstract, symbolic world created by the theories he regarded as established.[34]

Thus the efficacy of the senses as reliable receptors and conduits of material experience, capable of un-problematically yielding up scientifically useful observations, cannot simply be assumed or proved from within a logical positivist account of natural science (you cannot prove by observation, that observations are reliable). That is the dilemma facing the logical positivist. However once we stop playing the logical positivist game, and admit that there is no clear cut distinction between observation and theory (all observations are theory laden), we are then free to construct a spectrum of observability, which ranges from observations which are little more than sense data (again it is not even clear what 'sense data' or Fiegl's 'raw feels'[35] are – are they a neurological response to outside stimuli – in which case they are utterly private and unique, and generally inaccessible even to ourselves, or are they statements like – 'I see a red patch in my visual field'? – statements expressible and communicable within human language, and liable, apart from anything else, to distortion from a whole variety of causes, from hallucinations and brain damage to poor lighting[36] – Feyerabend never offers us such details) with minimal theoretical components and very high degrees of inter-subjectivity, to highly theoretical constructions, such as the observations of quark pairs in large hadron colliders, understood only by specialists in the field, and yet still having possible empirical falsifiers.

Such a 'continuum of observability'[37] will enable a more sophisticated and less dichotomous approach to theory and observation. It will also enable us to construct a corresponding continuum between theory-laden and theory-determined observations, which demands that in each case we investigate the nature and degree to which theory informs and frames observations. Torrance here calls for a disciplined, intuitive and dynamic co-ordination of observational and theoretic frameworks – 'integration of form' – within natural and theological science.[38] Further on we shall understand that requirement as part of the hermeneutical component of all acts of understanding, even the natural scientific.

---

[34] Duhem, 1954, p. 159.
[35] Hahn & Schlipp (eds.), 1998, p. 320.
[36] Bacon, 1994, p. 22.
[37] Moore, 2003, p. 65.
[38] Colyer, 2001a, p. 337.

Neither does the distinction between theory and observation necessarily have to be understood simply in terms of terms and statements. "Certainly empirical research can never be a-theoretical, but it would seem to be a different *activity* from theoretical debate."[39] The italics are mine – terms, even statements, may be identical but their meaning will be only fully determined with reference to their context. Feyerabend admits as much:

> There is of course a distinction between theoretical terms and observation terms, but it is a *psychological* distinction, dealing with psychological processes that accompany their use, but having nothing to do with their content.[40]

This admission of Feyerabend's is I believe more important than he makes out. I would suggest – to turn Wittgenstein on Feyerabend, we might say that there is a readily identifiable distinction between the language games of observation and theory, even when the terms employed are the same.

Feyerabend is here then, keen to press us, somewhat prematurely, to radical conclusions, because a more measured approach (such as Torrance's) to the theory / observation question, would not lead us on the road to an anarchist epistemology, the next step of which is counter-inductivism and theoretical pluralism. Theory *laden* observations demand that strategic 'hermeneutical' decontamination takes place, theory *determined* observations justify us in positing novel theories which fly in the face of observable evidence – what Feyerabend labels 'theoretical pluralism'.

This counter-inductivism[41] is a challenge to the demands of empirical cumulativism. Empirical cumulativism[42] is the conservative insistence that any newly proposed theory must account for the empirical data of previous theories, indeed it must preferably expand it by accounting for anomalies within past theories and producing novel predictions. Cumulativism is also closely related to reductionism, for hopefully the new theory will be successful in showing that the old

[39] Sayer in Archer et al. (eds.), 1998, p. 122.
[40] Feyerabend, RRSM, p. x.
[41] Feyerabend, AM, p. 68 – lack of empirical cumulativism is however often tolerated at the early stages of theory development, especially if the theory has some other very desirable features – but counter-inductivism *as a strategy* is Feyerabend's peculiarly radical version of this.
[42] Preston, 1997, p. 86; Scheffler, 1982, p. 9.

theory was simply a 'special case' of the new one. Feyerabend insists, to the contrary, that new theories need not be as empirically adequate as old ones, even if that leaves us with no unrefuted theories. Galileo is his prime example,[43] and it is certainly true that *at the time* Galileo did not have the weight of empirical evidence on his side – much of this came later, in some cases centuries later.[44] The heart of Feyerabend's anarchy resides here – for if we cannot decide between a well established and verified theory on one hand, and *any* theory – however outlandish – on the basis of the evidence, then as Feyerabend would say 'anything goes'. Whether we would wish natural science as cultural activity to be reduced and humbled in this way, thus becomes a sociological *decision*, and whilst Feyerabend admits that anarchy is not a desirable situation on a permanent basis, it can be effective medicine in states where oppressive epistemic regimes need to offer legitimate freedoms.

This theoretical pluralism supposedly explains Feyerabend's enthusiasm for Einstein, an enthusiasm shared, but for rather different reasons, with Torrance. Feyerabend's Einstein is a rather different Einstein from Torrance's. Feyerabend's view which emphasises the initially speculative quality of Einstein's theory is based upon the supposedly meagre empirical support enjoyed by that theory, and notably Feyerabend's assertion that Einstein's greatest qualities were – somewhat against the historical evidence[45] – his 'creativity' and "a strong faith, a deeply religious attitude."[46] As McGrath notes general relativity theory did not receive universally accepted empirical backing until the 1960s.[47] In particular Popper's assertion that "the Michelson-Morley experiment led to the theory of relativity"[48] often quoted out of context in textbooks, is disputed by both Feyerabend (1999) and Polanyi (1998, p. 9f), who make clear that the situation was more fluid and complex than that. It is of course a mistake to cite Popper out of context in this way, because as a falsificationist it is unlikely that he would believe that a particular experiment could 'lead' to such a large scale theory in

[43] Feyerabend, AM, p. 121.
[44] See footnote 111 in chapter two.
[45] Einstein, 1934, p. 11.
[46] Feyerabend, FTR, p. 133.
[47] McGrath, 2001, p. 239.
[48] 1992, p. 108, see also for instance Weizsäcker, 1952, p. 68.

anything like the mechanical way subsequently suggested. Certainly reception of the theory was not quick or universal.[49]

However as a justification of theological pluralism this is a controversial example and interpretation of history, especially when Feyerabend goes on to generalise this conclusion: "there is not a single theory which is in agreement with all the facts in its domain. And here I am not speaking about rumours or about the results of sloppy procedures."[50] Although Laudan endorses this view,[51] Penrose (and others) maintain that "no confirmed observations exist that contradict Einstein's relativity."[52] General relativity theory therefore triumphed despite *few* empirical confirmations and our inability to experimentally test its most novel predictions, but what Feyerabend is trying to suggest (to support his thesis of counter-inductivism) is that it has succeeded in spite of *opposing* empirical results. This is simply untrue. Although Einstein was indeed a creative and speculative thinker, he was (as we shall see in the next sub-section) also a very disciplined and realist one, and his genius, as for many great scientists, lay in the tension of that combination of attributes. Feyerabend's attempt to fashion him into a proto-anarchist, rather like his revisionist views of Galileo (though here he is on better ground), are historically one-sided.

Thus Feyerabend insists that the only imperative operative within theory creation should be that of human pleasure, and the only limits, the humanitarian boundaries imposed by his liberal socio-political outlook.[53] Whilst it is right to reject any mechanical algorithmic approach to theory creation, and emphasize the tacit and personal elements of theoretical reflection, one must also reject this more radical hypothesis of Feyerabend, whereby theoreticians are so completely relieved of the burden of having to account for the evidence.

However this seems a rather unlikely and perhaps undesirable means by which theories are created, chosen, and assume dominance within science, so we need to investigate further this important scientific process.

Logical positivism granted ontic priority to observation data given by our senses, this being sealed off from theoretical metaphysical

---

[49] Feyerabend, NSR, p. 162; Popper, 1969, p. 255. See also: Feyerabend & Lakatos, 1999, p. 100.
[50] Feyerabend, RRSM, p. 146.
[51] Laudan, 1977, p. 27.
[52] Penrose, 1989, p. 273.
[53] Feyerabend, AM, p. 183.

statements which were granted such reference only subject to their verification by such data. Theoretical statements therefore also enjoyed a highly contingent relation to reality, and might contain within them elements which were purely constructed or speculative. Feyerabend by breaking the seal, shows that all scientific statements – observational and theoretical – may be infected with speculative or constructed components. Thus by positive and negative feedback loops our recording of observational data will be heavily influenced by our theoretical framework, and the likelihood is thereby increased that theories will have large elements of construction, contingency and speculativeness within them. This is not a problem for Feyerabend – as long as we give up the idea that there is something special about scientific statements which exempts them from this normal state of affairs, and instead chose theories simply on the grounds that they promote human well-being.

Torrance seeks to draw the loop in the opposite direction – as an inward spiral, with the data of (theological or natural) science exerting a disciplining pull upon our theoretical frameworks, coercing them – albeit in a circling manner – towards increased verisimilitude, like a meteor caught within a gravitational field.[54] The extent to which theories are and have been speculative constructs within both theological and natural science is therefore of vital importance.

Feyerabend relies at this point[55] on a revisionist view of Mach[56] – surprisingly, as most regard him as the arch-exponent of logical positivism[57] – especially with regard to his non-realist epistemology,[58] though Einstein does admit a debt to Mach's forward-looking wish to revise Kantian notions of space.[59] Feyerabend attempts to separate Mach's epistemology from his 'physical arguments', and suggests that in fact it was 'lingering positivism' in Planck and Einstein which forced upon them an 'intrusion of religion'.[60]

With regard to theological science, the historically particular revelatory events of incarnation cannot be harvested and milled for their

[54] To push the musical metaphor of feedback loops, we note the avant-garde musicians of the 70s who *harnessed* the 'howl round' of the feedback loop for more creative ends.

[55] Feyerabend, FTR, p. 192ff.

[56] Hentschel, 1985.

[57] E.g. Bhaskar in Archer et al., 1998; McGrath, 2002, p. 139f.

[58] Torrance, 2001, p. 43.

[59] Einstein, 1993, p. 146.

[60] Feyerabend, FTR, p. 211.

timeless truths of reason, and the historical husk of particularity discarded, the process Barth calls "the nadir of the modern misunderstanding of the Bible."[61] Rather in faithfulness to object, theology must indwell the unitary rationality of salvation history, respecting the particularities and singularities of that history, as part of the fabric of time – time according to Einstein being itself a part of the fabric of reality and not the loom on which that fabric is weaved. Thus the importance of creation *ex nihilo* for Torrance in which "time is in creation, creation is not in time."[62] Thus if the theoretical statements (in chapter six excursus on hermeneutics we shall discuss what are the equivalents of theories and data for theological science) of theology are to be dogmatic statements, then we cannot and should not assume that ultimately those dogmas will consist entirely of statements which have no chronological specificity, as no such abstraction is possible.

Particularly since Einstein there has been an increased interest within physics with regard to the process of theory creation. Old inductive ideas, regardless of the logical problems of justification by induction, had not adequately addressed where theories were derived from, and how they were to be correlated to empirical data. Einstein's theory of relativity seemed an odd mixture of the highly speculative[63] with the empirically fruitful – for instance in its 'solution' of the perihelion of Mercury.[64]

Empirical data always underdetermines theory, because it is a mathematical triviality to show that however much data one has, it being finite, it is always *possible* to create an infinite number of mathematical relations between the variables involved.[65] Although that possibility was once thoroughly constrained by the relatively small amount of mathematical complexity with which the human brain can deal, but super-computers and advanced methods of numerical analysis have meant that such a possibility is now more of an actuality. In practise

[61] Barth, CD I/1, p. 329.

[62] Torrance, 1969a, p. 11 – McGrath, 2001, p. 168, defends the view that such a doctrine is specifically Christian and only later adopted within Judaism – this he notes is against the current consensus (against for example Williams). See also Barth, CD II/1, p. 608.

[63] Popper, 1969, p. 255; Einstein, 1934, p. 48.

[64] Relativity theory accounted for at least some of the already long known 43″ of arc which cannot be accounted for by classical mechanics, from the total 5600″ per century discrepancy from an elliptical path in the orbit of Mercury.

[65] Feyerabend, RRSM, p. 58; also Popper, 1992, p. 138; Polanyi, 1946, p. 9.

science assumes that solutions will be relatively simple and often simplicity will become a strong criterion of theory choice – no one would prefer a 184th order differentiation to a second order one unless the empirical fit was markedly better. However logically, no crucial experiment can decide between an infinite number of theories, simply on the grounds of empirical fit.[66] However, as we have seen, neither can empirical data achieve the weaker task of decisively falsifying a theory, as Popper would wish, because we can never be absolutely sure in ruling out any error in experimental design or calculation, and also more important logically, the Duhem-Quine underdetermination thesis rules out any attempt to isolate particular theories for testing.[67] It is however possible for a scientific community intersubjectively to elect a 'theory in question' and use the techniques of experimental isolation, control and variation in initial conditions, to make such election more transparent – but Feyerabend interprets that as that same 'community' simply closing ranks, perhaps sacrificing a relatively minor theory in order to save the 'mother-paradigm' on which the epistemic kudos of that community depends.

This inability logically to derive theoretical statements from observational data, so often thought to be an unproblematic advantage of the 'scientific' method, led Einstein to conclude that: "there is no logical bridge between phenomena and their theoretical principles."[68] There is no algorithmic, systematic or deductive method by which theory creation or decision can be carried out.[69,70] "Einstein showed

---

[66] The subset of solutions would still be infinite – another triviality.

[67] Lakatos, 1978, p. 3 & Duhem, 1954, p. 187.

[68] Einstein, 1934, p. 4.

[69] Kuhn, 1970, p. 200.

[70] Popper cites this rather naïve inductive view of theory creation, as one way in which some historians, not to mention theologians, tend towards a rather simplistic scientism:

"But I should go even further and accuse at least some professional historians of 'scientism': of trying to copy the method of natural science, *not as it actually is,* but as it is wrongly alleged to be. This alleged but non-existent method, is that of collecting observations and then 'drawing conclusions' from them. It is slavishly aped by some historians who believe that they can collect documentary evidence which, corresponding to the observations of natural science, forms the 'empirical basis' for their conclusions." (Popper, 1979, p. 186)

In the theological analogy of data and theory, it would amount to systematic theologians mechanically 'drawing conclusions' by stringing together Biblical texts, exactly the approach of Arius which Torrance understood Athanasius to be rejecting. (Moore, 2003, p. 132 – quoting R. P. C. Hanson, the Arians' "unreflective parroting of Scripture").

that while we do and must operate with a correlation between the theoretical and empirical components of science, we cannot derive the former from the latter by logico-abstractive processes."[71] Torrance however believes that Einstein is incorrect therefore to infer that "*nothing* more can be said" as to how concepts and experiences are to be correlated.[72] True rationality will here admit that the proper 'integration of forms' extends beyond logico-deductive procedures (Einstein's point), but this extra-logicality should not be interpreted as a relinquishment of rationality or scientific rigour, as per Feyerabend, but a disciplined dynamic integration using tacit and intuitive elements within personal knowing which Polanyi had so effectively catalogued as being part of all scientific procedure.

Torrance is clear that our concepts, however constructed, naturally interpreted and socially laden they are, can still progress towards an accurate depiction of the 'elemental forms' of which reality is ontically, albeit ontology as relation, constituted. Our experiences are experiences *of* something, at least putatively of reality, not pure esthesis (as though experiences are entirely 'our possession'), and therefore there can be no logical positivist relegation of theory to simply a short-hand for the observed. If we are prepared to admit, as seems reasonable, that our experiences have a degree of order which we are not conscious of imposing upon them, then one possible explanation for these ordered phenomena, is that there exists beyond us, an ordered reality governed by general natural laws which we might realistically aim to discern and elucidate – on what ground logical positivism rejects that as at least a highly likely hypothesis, is unclear.

However Torrance's realism should not be interpreted as arguing for a "direct correspondence between concepts and experience,"[73] but

---

[71] Torrance, 1985, p. 21.

[72] "The point at which I am challenged by Einstein is precisely that at which he insisted in his famous thesis on Physics and Reality that 'nothing can be said concerning the manner in which the concepts are to be made and connected, and how we are to coordinate them to the experiences'. I entirely agree with Einstein's refutation of Kantian and Machian ways of correlating the empirical and theoretical components in scientific knowledge, but if the scientist is a personal being, and not just a mathematical machine, then I believe that we must have rather more to say at this point. Of all our contemporaries no one has done more to advance our understanding here than Michael Polanyi, to whom I was not a little indebted in the last chapter, that is, to Polanyi's discussion of the tacit or inarticulate dimension in human thought as the area in which our spontaneous and creative thinking is at work in intimate and intuitive contact with reality." Torrance, 1985, p. 132.

[73] McGrath, 1999, p. 217.

rather to conclude as Bhaskar does that "theory is not an elliptical way of referring to experience, but a way of referring to the hypothesized inner structures of the world."[74]

As ever Feyerabend concurs in critiquing a polar position, but then rushes headlong to the other pole. So with regards to the importance of the speculative dimension within modern physics he remarks:

> It shows that the imposing and perhaps a little terrifying picture of science as an unalterable and steadily increasing collection of facts is nothing but a myth, and that ingenuity and speculation play in physics as great a role as anywhere else.[75]

What Feyerabend however ignores is that Einstein also draws attention to a strongly realist feature within physics, namely the fact that whilst there is the *possibility* of infinite theories which fit all data, in *actuality* it is more likely that the theoretical physicist will be searching for *one* which does – he or she will not be spoilt for choice.

> In this methodological uncertainty, one might suppose that there were any other number of possible systems of theoretical physics all with an equal amount to be said for them; and this opinion is no doubt correct, theoretically. But evolution has shown that at a single moment, out of all conceivable constructions, a single one has always proved itself absolutely superior to the rest. Nobody who has really gone deeply into the matter will deny that in practice the world of phenomena uniquely determines the theoretical system, in spite of the fact that there is no logical bridge between phenomena and their theoretical principles; this is what Leibniz described so happily as a 'pre-established harmony.'[76]

This is rather a blow for Feyerabend's theoretical pluralism and his appeal to Einstein for support, if it is to do anything more than alert us to a rather remote possibility. Of course Feyerabend's counter inductivism means that he would not worry about being adequate to theory-laden phenomena, but crucially that then loses him Einstein as an ally. Feyerabend[77] and Quine[78] draw our attention to the *possibilities*, but what is more remarkable is this 'pre-established harmony' – that is what scientifically cries out for explanation and fits well with the

[74] Bhaskar in Archer et al. (ed.), 1998, p. 59; see also Torrance, 1984, p. 273.
[75] Feyerabend, RRSM, p. 219.
[76] Einstein, 1934, p. 4; also Polkinghorne, 1998, p. 17 – "The problem is … not the selecting from a plethora of possible explanations but of finding one which is adequate."
[77] Feyerabend, RRSM, p. 59.
[78] Quine, 1990, p. 102.

Christian doctrine of creation. If that is an 'intrusion of religion', then it is hardly some imposition of a theological ideology, but a simple admission, as Torrance would say, that given the facts of how things are epistemically, here is an explanation, which happens to have a theological provenance, and which works rather well.

Thus crucially there is a distinction between what Feyerabend on the one hand, and Torrance and Einstein on the other, think that theoretical scientists will actually be engaged with. For Feyerabend the task of the theoretical scientist is to explicate her particular theory, chosen from several alternatives, chiefly on the criteria of human whim,[79] and to defend it tenaciously against the attacks of other scientists, especially if those attacks are mounted by establishment forces using the weaponry of their own theory-determined empirical data. Such artistic licence for Feyerabend, undermines all restrictions which might be imposed by method.[80] For Torrance and Einstein, the task of the theoretical scientist will be to construct or create one theory which at least approximates to most of the generally agreed data. Hacking describes this as the 'contingency' of the theory (though I think he errantly does not "attribute a strong contingency thesis to Feyerabend"[81]). At times the theoretical scientist will need to underpin the data with theoretic interpretations, and at others will need to devise theory-neutral crucial experiments to enable theory choice, but generally natural scientists will find themselves and their creative imaginations under the rather iron discipline of empirical adequacy. As Duhem puts it:

> Logic leaves the physicist who would like to make a choice of a hypothesis with a freedom that is almost absolute; but this absence of any guide or rule cannot embarrass him, for, in fact, the physicist does not choose the hypothesis on which he will base a theory; he does not choose it any more than a flower chooses the grain of pollen which will fertilize it; the flower contents itself with keeping its corolla wide open to the breeze or to the insect carrying the generative dust of the fruit.[82]

Whilst Feyerabend fails to enlist Einstein to the anarchist cause, Torrance also has a recruiting problem. The problem, for Torrance's realism, is that Einstein has little time for theology, at least of the sort

[79] What Preston (1998) calls Feyerabend's post-modern 'supermarket' conception of natural science.
[80] Feyerabend, AM, p. 167.
[81] Hacking, 2000, p. 98.
[82] Duhem, 1954, p. 256.

which Torrance endorses: "'Religious truth' conveys nothing clear to me at all." He goes on to describe his religious views as 'pantheistic'.[83]

When Einstein talks of 'possible systems of theoretical physics' he is referring to mathematically articulated systems of Western science – voodoo is not for him a 'possible system'. Neither however, is the dogmatic theology of any particular religious tradition. Therefore Torrance finds himself caught in the middle. His realist approach to natural science means that he looks to Einstein as a model, but when he tries, unlike Einstein, to expand his outlook to include dogmatic theology, Feyerabend is waiting ready to spring his trap, and ask why astrology, witchcraft and voodoo are not also to be included. Furthermore there are experts in the field, such as Abraham Pais, who now question whether Einstein's realism was characteristic of his later life, and that at the time of the relativity theories he was philosophically more eclectic and might equally well be regarded as a Platonist or a Pythagorean.[84] Both Feyerabend's and Torrance's appeal to Einstein finds him to be only a partial if prestigious ally.

Whilst Torrance is therefore correct to reject any mechanical algorithmic approach to theory creation, and emphasize the tacit and personal elements of theoretical reflection, on the other hand he must also reject the more radical alternative hypothesis of Feyerabend: that theory determines observation.

Feyerabend is here over-reacting in relation to positivist and Popperian views of science which make a sharp disjunction between theoretical and observation terms or statements. McGrath also denies this possibility of enforcing such a disjunction: "what was once deemed to be 'theoretical' can become 'observable'."[85] Some terms – for example 'gene' and 'electron' – which began as theoretical entities, have over time become observable ones. However, what are thereby initially referred to as 'unobservables' are still 'objects' of some sort or another, even if at first their existence is purely hypothetical. In this instance there exists a 'continuum of observability',[86] the place on that continuum being a function of how direct the phenomena which confirm the existence of that class of objects. Therefore putative objects which initially give rise to no detectable phenomena, may then give rise to

[83] Einstein, 1934, p. 11.
[84] Luoma, 2002, p. 71.
[85] McGrath 2002, Vol. 2, p. 150.
[86] Moore, 2003, p. 65.

indirect ones (orbital perturbations in a known planet) and thence to increasingly direct ones (telescopic sighting of a new moon) – for instance Van Fraassen, rather eccentrically, is a very strong empiricist, making the cut off point a way along this continuum, by, for instance, denying ontological reality to Jupiter's moons because they cannot be seen with the naked eye.[87] Sometimes there will be theoretical reasons why we cannot expect to move along this continuum, so that for instance we cannot expect to observe a quark in isolation.[88] However by theoretical, I am thinking more of 'non-object' terms such as 'energy', 'work', 'field', 'force', 'continuum', as Torrance is thinking of with his 'elemental forms'.

Whereas logical positivism posited a clearly delineated path from observation to theory, and whereas conversely Feyerabend imagines the two to be so intertwined as to become inseparably vicious, Torrance and Barth envisage, both within theological and natural science, a circular / iterative path which continually spirals between the poles of theory and data, with the realist aim of statements increasingly corresponding with proper rectitude to reality.

For Torrance, over time, the path of our knowing may in actuality spiral outwards as well as inwards, as scientific disciplines becomes lax, but the good scientist will orient herself towards the interior focus of our knowledge, towards the reality which is independent of our knowing, a reality which exerts a centripetal force upon us, which we, if we are properly disciplined, do not resist, though we may question whether other less scientific forces are operative, pushing us to premature conclusions. We spiral outwards when we ignore that pull, and instead find ourselves attracted by some other centrifugal distraction, which is likely to be manifest in humanity's unique and over-riding ability for self-fascination. Proper objectivity demands a conscious effort to reject such nominalistic self-fascination, and a re-orientation towards the reality which lies beyond us, and finds its ultimate basis in God. However there is a proper scientific centrifugalism which resists the simplistic centripetal pull of a naïve realism. This seeks to wrench us prematurely to convergence by a path on which we have not acted with due scientific caution towards inherited patterns of thought.

Torrance is insistent on the convergent possibility of this spiral, crucially, because whilst he, like Feyerabend, identifies certain circular

87 McGrath, 1998, p. 146.
88 Polkinghorne, 2007, p. 31.

features, Torrance also analyses the centripetal and centrifugal vectors which combine to form them, and seeks to promote the centripetal over the centrifugal, as opposed to Feyerabend, who simply points them out. According to Torrance relentless fidelity to observation can allow theory-laden observations to transcend the circularity of theory-determinedness, and spiral towards an accurate representation of the underlying reality. Similarly, theological data, by relentless questioning of its own implicit assumptions and metaphysical baggage, can seek to transcend its own circularity, and speak realistically and faithfully of God. In similar vein Habermas criticises Gadamer for underestimating "the power of reflection to criticize one's inherited prejudices."[89]

* * *

Whilst we might therefore reject the radical theory-ladenness of observations with which Feyerabend urges us towards a similarly radical theoretical pluralism, one feature of his account – the theory of natural interpretations – has genuine utility in giving a credible account of the mechanism by which data may come to be more resiliently theory-laden than expected, and in a way which is relevant within the field of theological science. Natural interpretations are "ideas so closely connected with observations that it needs a special effort to realize their existence and to determine their content."[90] Natural interpretations are not limited to natural science, for in all acts of human understanding, often imbedded within language itself, are prejudices (in the non-pejorative Gadamian sense) of which we are largely unaware.[91]

The danger of natural interpretations lies in their trivial and covert nature. They consist of the very many assumptions that we unconsciously[92] make as we perceive the material world. In this great slew of interpretative activity, so constant and yet nearly always carried out in the background of our consciousness, a great deal of immersing and conforming our perceptions to inherited noetic frameworks takes place, so that it is possible that some decisive and reality-forming interpretative decisions may slip through unexamined. Similarly many piecemeal fragments may coalesce, resulting in a set of perceptions which is remarkably dissimilar to that of other human beings.

[89] Schmidt, 2006, p. 142.
[90] Feyerabend, AM, p. 54.
[91] Schmidt, 2006, p. 101; Polanyi, 1998, p. 112.
[92] Feyerabend, TDK, p. 20.

To eliminate natural interpretations totally would be to eliminate "the ability to think and to perceive."[93] Even when we are able to bring such artefacts out into the open for examination, the Duhem-Quine thesis is still operative and it is likely that the web of mutual support will be spread over a wide variety of subsidiary material.

In particular Feyerabend and Torrance draw attention to what we might label sequential artefacts within natural interpretations: particular words or ideas, laden with metaphysical theory, which are not simply part of the existing or penultimate theoretical paradigm, but pass through several paradigm changes unnoticed, making it nearly impossible to spot their original theoretical associations. These slippery characters are therefore always likely not only to cross chronological divides but scientific ones, as the natural interpretations of chemistry (for example) produce distortive effects if carried over unquestioned into biology.

> Without any doubt the language that we use in physics or mathematics, for example, and certainly in theology, is already laden with metaphysical ideas, for it is shaped through centuries of history and culture and is impregnated with previously adopted attitudes to life or views of the world, but scientific procedure through sheer respect for objectivity must allow all this to be called into question.[94]

They are far more dangerous than more obvious systematic divergences between traditions, for "cultural assumptions, after all, are most dangerous when we are unaware of them."[95] The meaning of terms is given within a certain language game or context, and when we change that, so the meaning changes (even if the actual vocabulary does not), both within the natural sciences and certainly beyond them. Furthermore with natural language that change is continuous and inevitable, we cannot lay hold of our language long enough to disinfect it of the parasites it carries.

Thus for Torrance, and particularly in ideas adopted from Michael Polanyi, natural interpretations have their theological equivalents.

An important thing to note about natural interpretations is that they impinge upon the sort of encounters envisaged within the science / theology conflict hypothesis. They are like computer viruses or worms

[93] Grunfeld, 1976, p. 315.
[94] Torrance, 1969, p. 288.
[95] Torrance, 1980, p. 13.

which lie dormant within systematic structures. So for instance in the Thomist Biblical-Aristotelian synthesis, natural interpretations are not the obvious points of content where the two components of the synthesis are in need of reconciliation – e.g. the eternity of the world vs creation *ex nihilo* – for these conflicts are all too obvious. Rather they are philosophical artefacts and prejudices which whilst they do not directly contradict the Biblical account, commit the synthesiser to patterns of thought which may have effects at a later stage, when a more contemporary synthesis is attempted.[96] So Thomists are likely to find themselves, for 'religious' reasons, committed to defending styles or methods which are Aristotelian, when their secular adversaries are no longer bothered either way. This is a subtle form of scholasticism, and Torrance's best illustration is that of the Gnostics:

> Gnostics were sinning against sound method when they forced creation against its inherent pattern to change itself into types of things that have no real existence. In that way we can get any meaning we want out of nature or out of the Scriptures. The proper method is to attend to the arrangement of created things, in nature and in history, and to harmonise our thought with what actually exists or with reason, for in this way we operate with a rule (*regula*) derived from the nature of things. E.g. in subordinating creaturely realities to God, for that is in accordance with their nature, instead of subordinating God to them. This is what Iranaeus refers to frequently as the rule of truth.[97]

Torrance's most fully elaborated example of this phenomenon is in the assumptions about time and space ('container view') which classical science absorbed from classical philosophy.[98] Because liberal Protestantism was over-anxious to ingratiate itself with *this* natural scientific paradigm, both Barth and Einstein (and Planck) in their respective fields had to work hard to break through this Kantian neo-Platonic inheritance.[99] Only then could new paradigms emerge in both natural science and theology, and a new synthesis take place.

Given Feyerabend's interest in the relevance of linguistics for natural science, it is no surprise that he adopts from Whorf, the theory that such cultural baggage is hosted by language, a view shared with

[96] Patterson, 1999, p. 67.
[97] Torrance 1995, p. 106–7.
[98] Torrance, 1969a, p. 5 – from Pythagoreans and Platonists, less than Stoics, whose idea that bodies created space by 'making room' for themselves was more conducive to Christian theology.
[99] Torrance, 2001, p. 77.

Gadamer.[100] The Kantian 'intuition' of space-time is according to Whorf, not an intuition as such, but a linguistic artefact. Thus according to Whorf the Hopi word which we translate as 'time' has different associations, from the Western 'chronological container' word 'time' and its equivalent in other Indo-European languages.[101] As Van Fraassen says: "All our *language* is thoroughly theory infected."[102]

Ignoring for a moment Whorf's categorisation of Semitic languages under the Indo-European umbrella, it is therefore important for Torrance that Boman[103] identifies three key differences *imbedded within* the grammars of the Hebrew and Greek language respectively: the dynamic versus the static; the concrete versus the abstract; a unitary versus a dualist conception of the human. Thus Torrance's and Barth's insistence that as Hebraic ideas met Greek ones, they 'redeemed' metaphysical artefacts embedded within Greek language, even if they now used Greek vocabulary.[104] Certainly Harnack is right to say that some fathers, and subsequent commentators, worked in the opposite direction – shoe-horning the gospel into a Hellenistic conceptual framework – but at Chalcedon, and particularly Nicea, the former tendency was in the ascendant.[105]

The incommensurability thesis

Incommensurability is without doubt a potent idea, which more than any other has preserved Feyerabend's name, albeit as a maverick, in the canon of philosophy of science. Feyerabend's incommensurability thesis suggests that there is a logical 'disjunction' between various scientific paradigms or high level theories, meaning that their logical relations cannot be explicitly formulated, and that natural science will have to tolerate their co-existence (local instrumentalism) at least until further work might resolve the disjunction. The thesis was originally

---

[100] Gadamer, 1975, p. 442.

[101] Whorf, 1956, p. 153.

[102] Van Fraassen, 1980, p. 14, my italics; see also Schmidt, 2006, on Gadamer, p. 124.

[103] See Osborne, 2006, p. 78.

[104] Torrance, 1990, p. 198.

[105] See Marley, 1992, p 10; J. Macquarrie, *Foundation Documents of the Faith: III. The Chalcedonian Definition*. The Expository Times 1979, Vol. 3, p. 68–72, p. 72; also M. Werner & S. Brandon, *The Formation of Christian Dogma: An Historical Study of its Problem*, Harper and Brothers, New York, 1957.

formulated with reference to the divergent ontologies of relativity and quantum theory and those of classical mechanics, but latterly Feyerabend employs the thesis to defend the rights of traditions from well beyond the boundaries of Western natural science: for instance voodoo is a tradition incommensurate with that of Western technological medicine.

The first published version of his incommensurability thesis he dates to 1958,[106] and discussions of it from 1952. In these early stages he is considering the complementarity principle in quantum mechanics and also the relation of relativistic and classical physics. In *Explanation, reduction and empiricism* (1962) he introduced the notion of incommensurability by way of three characteristics:

1. The primitive descriptive terms of T cannot be defined in terms of those of T'
2. There are no 'bridge laws' connecting the two sets of primitive descriptive terms which are correct and assertable consistently with T
3. The principles of the conceptual apparatus of T are inconsistent with those of T'.[107]

In *Changing patterns of reconstruction* (1977) he said that incommensurability meant "deductive disjointedness and nothing else." At one point Feyerabend maintains that the thesis, as it appears in *Against Method,* was inspired by the unlikely trio: B. Snell's *Discovery of the mind*, H. Schafer's *Principles of Egyptian Art* and V. Ronchi's *Optics, The Science of Vision.*[108] However in *Against Method* itself its origins are stated to be Elizabeth Anscombe's work on Wittgenstein's later work, having close links to his contextual theory of meaning and the related Sapir-Whorf hypothesis.[109]

'Competing' high level theories lack the point to point isomorphic relations between respective ontological concepts and terms, which would allow us to determine their logical relations – i.e., that they contradict at certain points, or that one is reducible to the other, or that they do not overlap at all. Feyerabend's primary example of classical

---

[106] See AM, p. 211.
[107] Preston, 1997, p. 104.
[108] Feyerabend, KT, p. 140.
[109] AM, p. 259–60; Nielsen in Nielsen & Phillips, 2005, p. 149, also points out that a holophrastic strategy is available to counter the claim that incommensurability can be in some way derived from the Sapir-Whorf hypothesis.

and relativistic mechanics is notable in that this incommensurability is a diachronic one – he is comparing one theory which is generally held to have 'superseded' another within a single discipline – a notably *realist* approach. Indeed it only works as an example by being markedly more realist than the approach taken by most physicists, who believe that classical mechanics is simply a special case of relativistic mechanics (where speeds are negligibly low in comparison to the speed of light). Later on he will use the incommensurability thesis far more frequently to compare synchronic theories, often ones from widely dispersed cultural traditions – and thus will employ the thesis in a notably *non-realist* manner, where arguably, as we shall see, it simply doesn't work.

Feyerabend points out that terms such as 'mass' have different 'meanings' in classical and relativity theory. The term 'meaning' is ambiguous, and Feyerabend confuses sense and reference[110] – the point made by Scheffler to Kuhn in *Science and Subjectivity*.[111] Certainly mass has different properties in the two theories (whether as a quantity it is inertial-frame dependent) but whether that implies that their 'meaning' (as intuitively referring to '*quantitas materiae*') is incommensurable is unclear. For instance one can plot the logical and historical connections between the two sorts of mass and a simple equation gives their relation.[112] Polkinghorne is right to insist on 'charity of reference' if the scientific enterprise is to be allowed any sense of continuity, otherwise scaling down from paradigms, there could be no such thing as 'mistakes' and 'clarifications' within quite low level theories.[113] Kuhn's own version of the incommensurability thesis, as commentators are keen to point out,[114] demands that paradigm shifts occur only when there is sufficient *continuity* between successive paradigms as well as sufficient discontinuity – otherwise such shifts would mean the reconstruction of *everything*. As Waismann points out 'what a word means' is different from the 'meaning of a word', a tricky bit of grammar that forced Wittgenstein to re-evaluate the *Tractatus* schema, and it is here where Feyerabend fails to exercise due caution.[115]

---

[110] Sankey, 2007, p. 3f – semantic incommensurability is only a threat to realism if it deals with variation of reference rather than variation of sense.

[111] Scheffler, 1967, p. 54f – p. 58 "stability of meaning is required only in the sense of stability of reference in order to proceed without mishap."

[112] See Penrose, 1989, p. 282: $M = m/\sqrt{1-(v/c)^2}$.

[113] Polkinghorne, 1998, p. 105.

[114] Pfurtner in Küng and Tracy (eds.), 1989, p. 133.

[115] Waismann, 1965, p. 313.

A second example is in the attribution of colour to physical objects. Feyerabend[116] contends that whereas in non-relativistic physics colour is a property of an object, in relativistic physics it varies with the relative speed of observer and object. However one can hardly allow the addition of such a simple initial condition to make such a startling change to our theory of meaning. There is for instance a parallel change in the meaning of the word 'pitch' or frequency in classical mechanics with the discovery of the Doppler effect.

Much has been subsequently written about the thesis, though far more in the social than the physical sciences. The main problem to have dogged the thesis is establishing the sort of mistake involved simply in accepting both the incommensurable theories.[117] By definition there is no logical contradiction between them, and yet they must be in some way comparable (and involve reference to some sort of well-defined domain) to the extent that the comparison is uncomfortable. However the only really uncomfortable comparison is one which results in contradiction. The correct answer is that the theories propose, implicitly or explicitly, very different ontologies, but that is only a problem for strong realists, such as the early Feyerabend, and of course Torrance. However in this case there is then an obvious subsidiary metaphysical contradiction between those ontologies – catch 22. His early tactical use of local instrumentalism may give a hint of an answer – abandon realism and move towards global relativism.

Whether Feyerabend does indeed perform this volte-face suggested by his own thesis, remains a disputed issue. Farell and Sankey[118] believe that Preston,[119] Hoyningen-Huene, Oberheim and Andersen[120] are all wrong to interpret Feyerabend in this way, and that whilst he enjoyed a brief dalliance with non-realism around the time of *Science in a free society*, his view turned back to realism towards the end of his life (in *Conquest of Abundance*). My own view is that at the stage we are focussing on, Feyerabend's ethical voluntarism gives his thought highly non-realist tendencies. Furthermore pragmatically Feyerabend's challenge has sharper teeth if we put up a non-realist interpretation of

[116] See Feyerabend, RRSM, p. 30.

[117] Schaffler, 1982, p. 82; see also with reference to Kuhn, Shapere's review of Kuhn, 1969, in *Philosophical Review*, Vol. 73, p. 391.

[118] Sankey, 2007.

[119] Farell, 2001, p. 351ff.

[120] Hoyningen-Huene, Oberheim, Andersen, *On incommensurability*, Studies in History and Philosophy of science, Vol. 27/1, 1996, p. 131–41.

his work against Torrance's strong critical realism. Either way, with regard to incommensurability, one gets the feeling that Feyerabend is dancing around the realist circle simply to keep the thesis alive and meaningful.

Perhaps the most crucial point is that incommensurability is not the same as incomparability,[121] for as Feyerabend says, if all else fails, you can listen to two theories being recited over a soothing guitar to compare which is the most melodious[122] – not a terribly helpful suggestion. If incommensurability did amount to incomparability the problem would to a large extent disappear, for then as Newton-Smith, Preston[123] and Couvalis all point out, the truth value of each could not affect the other, though it might do in the future.

> Thus Feyerabend seems to have no adequate account of how it is possible for two theories to be incommensurable, but rivals. His only real argument is that his case studies have shown it simply to be the case that theories can be both.[124]

The real offence of this theoretical pluralism is against the law of simplicity: it is difficult to understand how we might psychologically operate holding as 'true' a plethora of theories, each with its own logical and ontic structure, and each relating to an identical domain. Ockham's razor, if nothing else, is in this case a psychological necessity.

Feyerabend is well aware of the difficulty, and Preston catalogues eight ways in which incommensurable theories, according to Feyerabend, can be compared. They are, according to Preston 'a mixed bag.'[125] Couvalis gives a slightly different account of how incommensurable theories can be compared, using the idea of a 'domain of investigation'.[126] The fact that incommensurable theories can and must be compared, is for Feyerabend a consequence of his early super-realism. Later on without that realism,[127] incommensurability slides

[121] Preston, 1997, p. 166 – Sankey, 2007, p. 1 makes this mistake: 'incommensurable suggests that the content of theories may not be directly compared' – in fact the opposite is true, incommensurability suggests that comparison is difficult but somehow demanded by the theories.

[122] Feyerabend, AM, p. 172.

[123] Preston 1997, p. 110 & p. 166.

[124] Couvalis, 1989, p. 90; also Moberg, 1979.

[125] Preston, 1997, p. 117.

[126] Couvalis, 1989, p. 99.

[127] Feyerabend, FTR, p. 73; Compare this with Rorty's similar pronouncement that "an objective public world [which] is not of our making. This formulae strikes me as no more than out-dated rhetoric." Rorty, 1991, p. 149.

into incomparability, and thence into relativism. However, though Newton-Smith observes: "both Kuhn and Feyerabend pass from the thesis of incommensurability to the thesis of the relativism of truth,"[128] Preston is right to point out, *that at its inception* "in fact, contrary to what critics like Putnam and Shapere have thought, the incommensurability thesis does not even entail relativism."[129]

Torrance's response would no doubt be that logical disjunction is not the only crisis which can engulf divergent ontologies, thus our choice between them will sometimes have to be made on extra-logical grounds – rationality demands it. Ultimately for Torrance, the realist, scientific rigour will push towards unitary frameworks, but as a critical realist he acknowledges that frameworks also need to be adapted to the traditioned and shifting contexts in which they find themselves, so that at times is would be premature to rush the choice, the integration of forms must be ordered and reasoned, if not always logical.

Against Method and Farewell to Reason

Theoretical pluralism must be distinguished from methodological pluralism, a later development in Feyerabend's thought which emerges as the core thesis of his 'epistemological anarchy' in *Against Method*. The latter however is a logical consequence of the former, if, as Feyerabend and Torrance both maintain, scientific methodology is to some extent dependent upon theory, rather than upon "an abstract view of rationality."[130]

In his correspondence with Imre Lakatos, Feyerabend locates the inspiration for his methodological pluralism in Popper's lectures:

> My 'position' now is thoroughly Popperian: Karl always starts his lectures on scientific method with the remark that 'there is no scientific method'. It is this remark which I accept – and I reject everything which comes afterwards.[131]

It was Popper's failure to abide by this opening gambit, which Feyerabend found so infuriating, and which Feyerabend insisted was

[128] Newton-Smith, 1990, p. 10, notwithstanding that Feyerabend himself insists that he uses the term incommensurability in a different sense to Kuhn.
[129] Preston, 1997, p. 121.
[130] Feyerabend, SFS, p. 99.
[131] Feyerabend in Feyerabend & Lakatos (eds.), 1999, p. 120.

taken at face value.[132] Indeed if the reminiscence is accurate[133] then the charge seems fair enough: Popperians indeed do not generally appear to be 'only a few rules of thumb' characters.[134]

Methodological pluralism for Feyerabend has a strong and a weak form. The weak form, consonant with Torrance's view, which is fairly conservative with regard to the scientific establishment, is that methods and techniques for each particular discipline and individual research programme are developed *in via*. However any scientific method will have a 'family resemblance' comprising qualities, which whilst somewhat overarching and vague, can be appreciated and to some degree assessed by all members of the scientific community: objectivity, precision, and structural coherence, for example.[135]

Feyerabend believes that this view, whilst gaining some crucial insights, is too timid. Strong methodological pluralism insists that methodological procedures are diverse to the extent that the scientific establishment is not competent to arbitrate such matters, which, because they often involve larger issues of what is taught to our children and how public money is spent, demand a wider system of accountability. Indeed Feyerabend understands 'scientists' to be simply a rather vociferous lobby group set on defending their own interests and that any appeal to 'objective assessment' and 'scientific method' are just propaganda terms that they have constructed for their own benefit.[136] 'Good method' should be what the whole of society judges to be for its own benefit rather than the ruling of an expert community delivered to a rather cowering public.

Torrance concurs with the weak form of methodological pluralism,[137] as does Ian Paul, writing in a series edited by Torrance:

---

[132] Feyerabend, FTR, p. 283.

[133] Repeated in KT, p. 88.

[134] Preston, 1997, p. 177.

[135] Polanyi is an important exponent of this view. See Feyerabend, KSR, p. 218; also p. 97 of this book.

[136] Feyerabend does not pull his punches: "Such illiterates (scientists) also determine where and how we are going to use nuclear power, how our children are going to live, what is good medicine and what is not, they waste millions of taxpayers money on ridiculous projects and get up in arms when a better control of these moneys is suggested", Feyerabend, TDK, p. 68, also p. 140.

"State Colleges and Universities are financed by taxpayers. They are therefore subjected to the judgement of the taxpayers and not to the judgement of the many intellectual parasites who live off public money." SFS, p. 134.

[137] See earlier quotation in footnote 98, chapter four.

> Most contemporary scientists also acknowledge that adherence to a single scientific method is a simplistic inhibitive misrepresentation. Scientific method is often described as if it were a set procedure. There is no such thing as *the* scientific method.[138]

However Polanyi, Torrance and Paul, all eschew its stronger form – indeed Feyerabend having created his 'stink bomb' was generally required to defend its stronger form alone.

Twelve years after the first edition of *Against Method*, Feyerabend published *Farewell to reason*. The titles of the two books merit a ⇔ between them – the scientist who is against method is (according to Feyerabend) therefore saying farewell to reason,[139] and vice versa.[140] "For Feyerabend, the core of rationalism is the thesis that there is something worth calling the scientific method, something unchanging which identifies all good science."[141] However (again according to Feyerabend) 'good science' – successful, resilient science which promotes human flourishing – is neither methodical nor rational, so we must make a choice, and the humane 'scientific' choice, at least as medicine, can only be according to Feyerabend: a dose of anarchy and irrationality. Therefore both 'rational' and 'methodical' have come to function as propagandist adjectives.[142] Like method, reason, according to Feyerabend, to be worthy of the name, must be capable of prior, extra and universal specification, rather than unfold immanently within science. The rationalist, says Feyerabend, is someone who "has an argument for every move that he makes," and "man does not live by argument alone,"[143] but such pronouncements are not going to worry the sophisticated critical rationalist of today. Torrance in particular is constantly stressing the extra-logical nature of rationality, a critical rationality which knows its limits and efficacy – Feyerabend can only conceive of rationality as logical deductive inferences following a predetermined and universal form. Feyerabend is up to his old tricks again – telling us what 'method' or 'rationality' must be and then demolishing the straw man thus created. Having divinised (see quote 145

---

[138] Paul, 1982, p. 16.

[139] "They can keep science, they can keep reason, they cannot keep both", AM, p. 214.

[140] Feyerabend, AM, p. 18.

[141] Preston, 1997, p. 170.

[142] Feyerabend, FTR, p. 10.

[143] Feyerabend, FTR, p. 303; Feyerabend in Feyerabend & Lakatos, 1999, p. 137.

following) rationality and methodology, he then delights in exposing them as mortal after all and tearing them down from Olympus, yet for those who have never bowed the knee this is simply theatre. The critical rationalist has nothing to fear.

Destination relativism

In his consideration of relativism Feyerabend takes us back to the pre-Socratics, to the 'regionalism' of Herodotus and the relativism of Protagoras.[144] Relativism is seen in terms of the political rights of minorities against the centralising tendencies of 'mythical' systems and authorities, and is more concerned with Herodotus' regard for custom, than Protagoras' reaction against the universalising rationalism of Plato's intellectualism. The following passage at the outset of *Farewell to Reason* is particularly telling:

> We may surmise that the idea is a leftover from times when important matters were run from a single centre, a king or a jealous god, supporting and giving authority to a single world view. And we may further surmise that Reason and Rationality are powers of a similar kind and are surrounded by the same aura as were gods, kings, tyrants, and their merciless laws. The content has evaporated; the aura remains and makes the powers survive.[145]

For Feyerabend, ideas of relativism and objectivism are not rooted in epistemology but in politics and theology.

Feyerabend's precise relation to relativism is difficult to penetrate, and often seems to be deliberately surrounded by his own obfuscation and indirectness. From the time of his incommensurability thesis onwards, it is always the philosophical abyss around which he skirts, until he does eventually cross that line, only to step back. It is doubly complex in that commentators maintain that Feyerabend has slipped into relativism before he has, and that those commentators include the later Feyerabend, who deliberately suppressed his own objections to relativism, only later on to suppress the suppression. It was however

[144] Feyerabend, FTR, p. 42f; He missed noting that Anaxagoras gives a good slogan of theory-determinedness, "things will really be for them however they believe" – see Aristotle, 1971, p. 18.

[145] Feyerabend, FTR, p. 11, chapter heading: *Notes on relativism*. See Cupitt, 1980, p. xi for a similar metaphor for religious faith.

always this charge which from 1962 onwards was liable to be laid against him (and Kuhn).[146]

Part of the problem is that many philosophers, Feyerabend amongst them, are inclined to make distinctions between different forms of relativism, which whilst often illuminating in themselves are incommensurable with other taxonomies. Newton-Smith for instance argues that trivial semantic relativism is the only relativism which is not self-refuting, whilst Bhaskar continually highlights the difference between judgemental relativism and epistemic relativism. Feyerabend himself distinguishes between philosophical relativism – which he endorses in *Farewell to Reason*, only to immediately repudiate in *Three Dialogues on Knowledge* – and political relativism which he endorses certainly from 1975, if not before, as well as using terms such as 'ontological' (from Quine), 'cosmological' and 'democratic' relativism. These distinctions are made explicit in *Science in a Free Society*. The key to the ambiguity is that for Feyerabend relativism is not a singularly well-defined concept, but one contrasted with objectivism and resting upon the false Platonic 'third world' assumption, that "concepts are stable and clear."[147] A less flattering assessment is in Hacking,[148] who identifies a total of thirteen versions of relativism within the Feyerabendian canon – "an attempt at divide and rule which fooled no one."

For what Feyerabend therefore avoids, is taking (for instance) Luggs'[149] critique of philosophical relativism by the horns. Instead, by numerous examples, innuendoes and diversions he is constantly endeavouring to argue from diversity to relativism, which is ultimately simply a *non sequitur*. Moore (arguing with regard to moral relativism) aptly names such a procedure a 'conjuring trick' and 'impossible'.[150]

Rather than conflating political, ethical and epistemological concerns as Feyerabend does,[151] one must in fact carefully divide them.

[146] "Relativism in philosophy of science, is the view that different scientific theories (or paradigms, or styles of theorizing) cannot be assessed by a single set of truth relevant criteria. But, as we have just seen, some of the criteria with which scientists can compare incommensurable theories are relevant to assessing the truth of the theories ...That Feyerabend was a relativist, knowingly or not, during this phase of his work (early-mid 1960's)...This is a myth ... then reinforced when Feyerabend himself finally did succumb to relativism." Preston, 1997, p. 121–2.

[147] Preston in Preston, Munévar, Lamb eds, 2000, p. 93 – again the question here is whether a stability of reference can be rescued from instability of sense.

[148] Hacking, 2000, p. 4.

[149] See Postscript on relativism in AM, p. 268.

[150] Tambiah, 1990, p. 128.

[151] AM, p. 14.

The political and ethical rights accrued by traditions can and must be asserted independently of their epistemic value. Otherwise, and contrary to Feyerabend's desires, legitimate freedoms will be in danger of requiring epistemic warrant. Astrologers must be defended from oppression, because they are (generally pretty harmless) human beings, not because astrology has passed some threshold of epistemic merit.[152]

The perennial problem with such espousals of relativism is the self-refuting notion that the subject, in this case Feyerabend, *has* however, Archimedes like, risen above the quagmire of competing view-points, and has ironically presented an account of 'how it really is'. Feyerabend's defence (one also used by Rorty[153]) is that in fact this is not what he has done. In so far as he has employed the 'techniques' of irony, inconsistency, dialogue, description, and counterpoint, he has not *asserted* relativism as another system, but has rather presented it as a pragmatic device, and left ultimate conclusions hanging. Critics might therefore accuse him not so much of self-refutation but of a hedging of bets – 'divide and rule'.

Voluntarism

At the interstices of ethical and epistemological concerns lies Feyerabend's voluntarism, the belief that ultimately we choose what sort of knowledge we shall have,[154] and therefore that choice must be made on ethical grounds.[155] Political interests and propaganda are not only present in science,[156] they are decisive.[157] Whereas Polanyi, upon encountering historical examples where this has occurred,[158] prescribes preventative medicine – Feyerabend's therapy is containment: admit the universal role of choice and desire, and promote good humane ones.

[152] Though of course it has been argued that in terms of empirical verifications astrology is historically the best served – Lakatos, 1978, p. 2; Popper, 1969, p. 256; Polanyi, 1998, p. 168.

[153] Notably, according to Rorty the pragmatist does not so much possess a *different* theory of truth, but rather has *no* theory of truth. Rorty singles out certain "incautious passages in the early writings of Feyerabend" (Rorty, 1991, p. 25) as being self-refuting in this way – presumably 'early' means the time in which Feyerabend was still an enthusiast of realism.

[154] See Feyerabend, RRSM, p. 19.

[155] See Preston, 1997, p. 20–21.

[156] Tambiah, 1990, p. 142.

[157] Lauden, 1977, p. 4.

[158] Polanyi, 1998, p. 52, 274 – examples are Mesmer and Elliotson.

This is a, perhaps the, crucial point of divergence between Torrance and Feyerabend, for Torrance's realism constantly reinforces his idea that knowledge is ultimately, despite the 'personal coefficient', not something we choose, but something which the object of study dictates to us.[159] If Feyerabend is right about voluntarism then the challenge to Torrance's realism will be devastating.[160]

Feyerabend does not shy away from the more unpalatable consequences of such a thesis: "thousands of women lost their breasts when they could have been healed by simple massage, diet, acupuncture, herbal treatment."[161] The question Feyerabend does however shy away from is who is responsible for that unfortunate decision: doctors, public health funding authorities, or indeed the women themselves?

Such voluntarist views, McGrath[162] notes, emerge later in the writings of radical social scientists such as Latour, Bernal, and Mannheim[163] – particularly Latour's rendering of Einstein's relativity theory as a purely social construction – and these views (which verge on the ridiculous) have received the dismissal they have deserved from better informed philosophers of science such as Ian Hacking and Alexander Bird.[164] Indeed Hacking's excellent *The social construction of what?* is by no means hostile to the idea of social construction *per se*, only to social

---

[159] Recollection of answer to interview question at Princeton quoted in McGrath, 1999, p. 57–8.

This is not to say either that Torrance is a naïve realist who believes that the knowledge of reality can be abstracted from acts of human knowing, complete with their personal, imaginative and speculative components, or that conversely Feyerabend believes that we simply choose what we want to know irrespective of the prevailing physical reality. Rather both are quite close together around the centre of this spectrum – it is simply that *ultimately* Feyerabend insists that knowledge is a matter of choice, and that Torrance insists that conversely it is the force of a reality beyond our choices and decisions – they fall on different sides of the razor.

[160] As noted earlier this is to admit an interpretation of Feyerabend that is controversial (i.e. we are siding with Preston against Farrell). In one sense this offers a pragmatic advantage in that Feyerabend thereby offers a clearer challenge test to Torrance's theological science, but I also make the case that at AM and for most of next decade this was Feyerabend's position.

[161] Feyerabend, SFS, p. 157.

[162] McGrath, 2002, p. 186f.

[163] But without the irony or the detailed knowledge of natural science which Feyerabend displays.

[164] "Many science haters and know-nothings have latched onto constructivism as vindicating their impotent hostility to the sciences" – Hacking, 2000, p. 58.

Bird is an interesting case, in that he declares that such social constructivism is not defended by any 'philosopher of science of note', and traces the inspiration for such excesses to Kuhn – it is unclear and intriguing whether he is unfamiliar with Feyerabend's influence, or simply does not count him worthy 'of note'. Bird in Moore & Scott (eds.), 2007, p. 62.

construction as global creed – it could equally well be entitled *The Social Construction of what and how much, and what does it matter?* Certainly our ideas about anything (an A-Z list forms the structure of the book), reinforced by the positive feedback which then determines our prejudices concerning any novel object, are socially constructed, but the important interpretative dynamic is in the detailed analysis of how such ideas are formed and what extrinsic references they possess, not a lazily disjunctive assertion of 'construction'. As D. Z. Phillips points out with reference to Durkheim, such reductionism is the result of a simple logical inversion: "instead of saying that our values are our bonds, he says that our bonds are our values."[165]

For Feyerabend it was ultimately a specific political event relating to the educational policy of the state of California which focused his views, rather than an ideational one. This underlines the relative importance of sociological and epistemological factors for him:

> In the years around 1964 Mexicans, blacks, Indians entered the university as a result of new educational policies. There they sat, partly curious, partly disdainful, partly simply confused, hoping to get an 'education'. What an opportunity for a prophet in search of a following! What an opportunity, my rationalist friends told me, to contribute to the spreading of reason and the improvement of mankind! What a marvellous opportunity for a new wave of enlightenment! I felt very differently. For it dawned on me that the intricate arguments and wonderful stories I had so far told to my more or less sophisticated audience might just be dreams, reflections of the conceit of a small group who had succeeded in enslaving everyone else with their ideas. Who was I to tell these people what and how to think?[166]

* * *

Feyerabend's political and ethical viewpoints would be peripheral to our epistemological concerns, were it not for the fact that Feyerabend is constantly trying to conflate them. He does not argue purely in epistemological terms but by constantly pointing out the political and ethical implications of particular epistemological viewpoints. It is my contention that this conflation is a strategy of avoidance – Feyerabend is constantly pointing out these ethical and political implications because he is trying to elicit our liberal humanitarian sympathies as

[165] Phillips, 2001, p. 239.
[166] Feyerabend, AM, p. 263.

cover in order to support a suspect epistemological relativism. However Feyerabend himself foresees the epistemic difficulties which such a relativism entails, and his actual embrace of it is distinctly guarded and qualified – a caution which many later writers forego.

A little context is important here. Feyerabend is writing in reaction to the demise of logical positivism and the rather strait laced falsificationism of Popper – when scientistic ideology was at the zenith of its popular appeal. Arguably he over-compensates in response to these (particularly with regard to Popper which most definitely 'gets personal'[167]), which in turn makes it difficult to afford his writings neutral assessment. He is writing at the outset of post-modernism (beginning to question the Enlightenment priority given to the meta-narrative of natural science), and in the past thirty years policy makers have arguably moved in a Feyerabendian direction with regard to the place of science in society[168] – and the 'radical' ground has also shifted correspondingly. Although an increasing proportion of funding for scientific research now comes from the private sector rather than government,[169] the increased variety of commercial patronage entails, for instance, a rise of interest in and use of complementary[170] and alternative medicines, and greater recognition of the vested interests within supposedly 'objective' scientific research, for instance with regard to the debate about climate change.

Conversely there has been increased recognition that in a democratic multi-cultural society diverse traditions accrue rights (and even the right to government funding) purely by weight of numbers, and not through any prior assessment of their epistemic merit. However whilst the post-modern distrust of meta-narrative, including natural scientific meta-narrative,[171] has gained ground with regard to epistemology, it has met more resistance than Feyerabend's political and ethical agenda. Thus the epistemological pole of Feyerabend's

[167] E.g. Feyerabend, TDK, p. 50; KT, p. 109.

[168] Hacking, 1975, p. 128, calls Feyerabend "the Cassius of the present plot … Davidson … its Brutus".

[169] Given that (if Feyerabend is right) share-holders profits and armament production are therefore going to be the concerns 'embodied' in such research, he would be uneasy about such developments. See also Hacking, 2000, Ch. 6, p. 163ff. Also Tina Beattie, 2007, p. 95, for figures regarding increases in military funding of scientific research.

[170] The title implies Feyerabend's notion of incommensurability.

[171] Often however such distrust is in a particular limited domain – few offer *alternative* narratives to those of quantum mechanics and relativity theory.

conflation is in fact *weakened* by the unequal progress made by his political and ethical concerns, by those who admit the former *without* a corresponding epistemic shift.

Feyerabend's epistemic destination is relativism, but that destination is fraught with dangers, logical and typological. It is a destination reached by his incommensurability thesis, whose notoriety stems more from its subsequent widespread adoption in the social sciences[172] than its influence within natural science. The problem with this thesis, is that when it is made clear precisely what other philosophical commitments are held in tandem with it, particularly with regards to realism, it is likely to turn out either to be innocuous or false.

[172] Preston, 1997, p. 102.

CHAPTER SIX

TWO EXCURSES

Having explored something of the relevant thinking by Torrance and Feyerabend we are nearly ready to deal with the cores issues that divide them: coherence, language and foundationalism. However in order to do so we shall need to make two brief excurses into related fields in order to obtain necessary tools.

For Torrance, theo-logic is a type of reasoning that is concerned principally with the logic of word – in particular the reading, reflection upon, and interpretation of Biblical texts. For some that will immediately suggest that theology cannot be a science, because a science must deal principally with the logics and quantity and number. We shall seek to understand how hermeneutics might be used as a scientific tool, and in particular how Torrance attempts that task. Furthermore we interpret Feyerabend as advocating a 'hermeneutic of suspicion' as the proper way in which science should be 'read' – after all says Feyerabend, scientists are really just creating works of art!

Secondly the question of how any text – be it a dogmatic statement, a Biblical text, a scientific theory, a mathematical equation – relates and maps onto an 'external reality' – if indeed such a thing can be talked about at all without contradiction and confusion – has been much discussed within both scientific and theological circles. Though there are similarities in the two discussions, there are also fundamental differences, and it is necessary to try and unpick some of the sometimes rather fine nuances and distinctions internal to these discussions, if one is to avoid drawing false or imprecise conclusions.

Hermeneutics[1] and science

If we allow that the data (see following) of the dogmatic theologian is the Biblical witness, the task of that theologian is permeated with hermeneutical concerns. Torrance rightly contends that the nature

[1] We here refer to 'hermeneutics' in a general sense, as Jeanrond (1994, p. 1) or Schmidt (2006, p. 1) uses the term as 'the theory of interpretation', and not as a specific type or school of interpretative procedure.

of the object of theology dictates, in faithfulness to that object, that we analyse the relevant data by chiefly employing the rationality of word. If however science is demarcated by our ability to render the data of each science in numerical, symbolic or organismic[2] form, then theological science, not alone, must relinquish its claim to be thus predicated. Whilst, where appropriate, the rationality of number enables us to employ powerful and elegant techniques in data manipulation and theoretical explication, and may render conveniently binary methods of verification, so that wherever quantifiability can be used we should seize its benefits, it should not be employed beyond its boundaries of applicability. Furthermore we contend that the discernment of that applicability is itself a task requiring hermeneutical instruments, so that all the natural sciences, even physics, contain such a component, though we fully admit that the size of that component will vary considerably. Hermeneutics[3] then becomes present in all acquisition of knowledge, and is furthermore fundamental for theology.

Feyerabend radicalises this notion by contending that natural scientists are simply creating works of art.[4] Presumably by this he means that the presence of ambiguous and tacit components within all branches of science, which necessitate full use of the human imagination, indicate an understanding of science which moves beyond the practical and mechanical towards the 'artistic'.

Whilst it is beyond the scope of this book to examine the nature of artistic depiction of the object, it should be made clear that here Feyerabend's reaction to realism and subsequent voluntarism has driven him to a false statement. For whilst scientific speculation does require a degree of creative freedom, scientists are bound by the nature and reality of that object – in some cases their theories can be confirmed by repeated and varied verifications of that object's properties - whereas the artist, often operates either without an object, or enjoys far greater degrees of freedom with reference to the reality of the object, so that judged as a *work of art* it is independent from any assessment of its first order correspondence to object – or as Ricoeur argues "in fiction and poetry the first-order references of the work are abolished."[5]

Whilst as a realist, Torrance understands scientific meanings as strongly bound up with notions of reference, Hacking maintains that

---

[2] McCabe, 2007, p. 140.
[3] 'The tacit component' – Polanyi, 1962, p. 69ff.
[4] Feyerabend, FTR, p. 294.
[5] Schmidt, 2006, p. 155.

Feyerabend's attack on foundationalism – as we have noted beginning with an attack on the theoretical / observation divide using the tools of linguistic analysis – is constitutive of an attack on 'meaning' itself.[6] Whilst he does not reference them, Feyerabend picks up something of the heady air of the post-structuralism and deconstructionalism of Derrida and Barthes, who both critique the notion of meaning as privileging a Western bourgeois culture which has captured and fixed the schemes of signification. However the alternative of a constant flux which never penetrates beyond the level of artistic play is inevitably hostile to any notion of scientific truth, and seems unnecessarily nihilistic – it is probable that natural science offers as much to the poor and marginalised of the world as deconstructionism. Pharmacies may be ambiguous places but antibiotics to combat infection have genuine utility.

* * *

Before proceeding further however, we need, more fully to establish and investigate an analogy previously alluded to. In order to understand the hermeneutical dynamic of theological science it is necessary to introduce a widely used analogy between natural and theological scientific thinking, which clarifies what we mean by data and theory.

Feyerabend's attack on foundationalism is focussed upon the data / theory dichotomy within natural science. Although he is initially sympathetic to the attempt by logical positivists to render any such dichotomy hermetic, his later work uses as a starting point a denial that any such attempt can be successful or indeed desirable. However for our purpose a rough distinction between observational data, and the theories which account for them, is a feature of natural science – even if only a psychological one – and any case for theology's equivalence of status will be greatly aided if a convincing analogue can be found. Torrance's analogue, widely shared particularly within Protestant theology, is that the Biblical material is analogous to the data of natural science, revelation is analogous to discovery,[7] and that dogmatic theology is the theoretical and systemising reflection on that data.[8]

---

[6] Hacking, 1975, p. 128; see also Waismann, 1965, p. 335.

[7] Torrance, 1969, p. 131.

[8] Torrance, 1996, p. 115. See also Osborne, 2006, p. 347; Moore also attributes this approach to Peacocke and Soskice, 2003, p. 44.

The differences between the poles of the analogies are determined by the differences of the objects of knowledge: dogmatics' object of knowledge being the event of revelation by God of himself in Jesus Christ. Alternatively Moore sees these differences as being significant enough to push the analogy towards disanalogy:

> An important disanalogy between theology and science arises concerning what each regards as 'data'. Whereas science depends on data whose availability is not in principle spatio-temporally unique and which can therefore be the subject of repeated experimentation, I wish to suggest that theology operates upon the 'data' or 'evidence' which are given by the scriptural witness to God's self-revelation. … Alister McGrath argues that theology and science have divergent epistemologies: 'The theologian is unable to appeal to present experimentation, or the results of past experimentation … Whereas the scientific community takes its ideas from such experimental approaches, the religious community takes them from revelation.'[9]

Although Torrance would concur that there exists a divergence over repeatability, this is caused by the personal nature of the object rather than its extra-sensibility, so that theological science is akin in this respect to the human sciences, rather than being completely isolated and concerned with an entirely supra-sensible noumena: "we are not concerned in our understanding of Christ with timeless and spaceless acts of God but with acts of God which fall within the space-time structures of human existence."[10] In that Jesus spoke and was heard, acted and was seen, touched others and was himself handled for good and evil, he belongs with us in history, which is grounded in (though not completely elucidated by) the physical world. This sense data, being personal and particular, is however therefore unrepeatable.[11] That is not as Moore contends an 'important disanalogy' demarking science from theology, for plenty of sciences are in the same situation – an example of the modern tendency to *think* 'physics' when we *say* 'science'. As we have seen, other sciences are concerned with personal / abductive data, just as the physical sciences are concerned with the impersonal inductive. Furthermore we have noted that 'evident to the *senses*' refers to something always private and unrepeatable, and its transformation into 'scientific data', even within physics, is by no means free from epistemic problems.

[9] Moore, 2003, p. 45, see also p. 50.
[10] Torrance, 1969, p. 312; see also Torrance, 1969, p. 208.
[11] Murphy, 1990, p. 53; Rae, 2005, p. 4.

Once the strict observation / theory dichotomy of logical positivism is dismantled however, we note that any impetus present within the analogy which pressured theologians into driving a wedge between systematic theology and Biblical studies[12] is reversed, so that the discipline of their spiralling inter-relation informs a scientific method which now finds itself deprived of those hermetic dualisms of theory / data, fact / value, science / ideology, subject / object, science / hermeneutics.[13]

Another important distinction within the analogy is the question of initiative.[14] In a natural scientific experiment it is the scientist who takes the initiative: he at a time and place of his choosing, sets up the experiment, controls the conditions, records and processes the data, and decides which subsequent lines of investigation to pursue. He decides in advance which questions he wishes to address to nature via the medium of experiment. He assumes that generally many of these decisions – such as the time and place of the experiment – will not affect the outcome of the experiment. Human sciences are a middle-point because persons as objects of study are able to interact with those decisions, making ones of their own, becoming agents of their own initiatives – perhaps choosing not to co-operate with the field worker.[15] In theological science, the initiative is utterly Godly, and nearly all these privileges are denied the scientist: "we have always to remember that we have to do with God Himself, and not with an object which we can clearly survey and control."[16] Fundamentally the data of theological science is given, rather than sought and obtained.

Our assessment of this analogy, on the positive side, is that it provides us with an observation / theory schema for theology, which has a high utility in any discussion of scientific method. On the negative side, the schema must be developed and particularized within each science, and we must ensure that we do not to incorporate disanalogous natural interpretations from sciences which have become improperly paradigmatic simply by their flamboyance. For example the requirement that observational data should be 'repeatable', cannot be perfectly maintained in any of the human sciences. Medieval arguments

---

12 Ricoeur in Küng and Tracy (eds.), 1989, p. 243.

13 Lamb in Küng and Tracy (eds.), 1989, p. 86.

14 Torrance, 1969, p. 98; Polanyi, 1998, p. 279.

15 Feyerabend, 1999, p. 137 – his characteristic examples are "shy birds, the easily bored and undercover agents."

16 Barth, CD II/1, p. 658.

over analogy, equivocity and univocity underline both the negative cautionary requirement that in speaking about the ultimate and unique object of God, our words reach the limits of their ability to reference, yet more positively, in the *lalia* of Jesus Christ, historical and concrete, our attempts to push those limits and speak truthfully of God are justified by God himself.

Torrance, coming from a quite particular ecclesiastical tradition, is working with a similarly particular schema concerning the relation of incarnation-revelation-scripture, which preferences the scriptures over more pneumatic[17] and ecclesial[18] sources of theological knowledge (see p. 236f). This is compounded by the perennial problem of this style of realist Biblical theology, namely that it must demonstrate – for it is not obvious – that there is *a* Biblical theology, rather than pluriform theologies within the canon (see p. 213). However here Torrance *has* learned a valuable trick from natural science – albeit a distinctively realist one. For natural science does not seek to prove, but use as a working assumption, the order and simplicity of the natural world[19] (an assumption which at the epistemic level Feyerabend contends is unfounded and imperious). Thus rather than *show* that the canon possesses a theological unity, Torrance assumes that below the human accretions, the canon, in a realist manner, witnesses to God whose unity and simplicity are foundational. Witnesses, within and without the Scriptures, need not be necessarily very good or convincing ones (we need simply to show that we do not have any obvious better ones) – the unity of their witness is one of reference, founded in the reality of the object to which they witness.

The dynamics of the analogy are effected by Torrance's realist conception of natural science, as opposed in particular to a logical positivist one: by analogy, the latter ideal 'dogmatician/theoretician' is minimally additive, being concerned merely with constructing instrumental rules secondary to the 'facts' of data, whereas the realist

---

[17] For balance we note Deddo in Colyer 2001 emphasises the importance of the Holy Spirit in Torrance as counter-balance for overly-rationalist methods in theology. The point however remains that Torrance is so strongly Nicene in his emphasis on the *homoousion* of the Spirit, the Spirit which mutually interpenetrates and mediates the Son, that his energy is devoted to contrasting this Spirit of God with any "confusion of the Spirit with the inward morality, spirituality or religiosity of humanity or the church." Colyer, 2002a, p. 212ff.

[18] Colyer, 2001a, p. 243ff.

[19] Torrance, 1985, p. 16.

dogmatician/theoretician is positively, actively and speculatively seeking to push right down to the elemental basis of their science.

The 'observation' language of Biblical studies, and the 'theoretical' language of systematic theology are implicated within one another, to such an extent that should a particular Biblical term or concept *not* find employment in systematic theology, the latter would be lacking; and similarly extra-Biblical terms and concepts introduced by systematic theology, though allowable, must be carefully used, disinfected as far as possible of natural interpretations, and utilised flexibly enough to allow the Biblical material to shape their reference and meaning – they must play by the rules of the Biblical language game.

It would be a serious error according to Torrance, if there were ever to be a serious rupture between these disciplines. At every stage the student of Biblical studies, in attempting to understand *what the Bible says* will have a subsidiary awareness of the theological framework of interpretation which the different churches have employed, and the different ways in which churches have understood individual texts. So too the systematic theologian, in systematizing the Biblical witness into a rigorous dogmatic framework, cannot simply study that framework, as it appears in its various historical guises, and note its inner connections, mistaken relations, and coherent implications, but yet fail to anchor that framework in its extrinsic correlations with the content of the Biblical text, with which it continually interacts.[20] This interaction is constantly iterative – just as Biblical studies can decisively inform doctrine, so doctrine can decisively inform the correct interpretation of individual texts. Neither can we 'finish' (i.e. extract the objective meaning) with the text (though barring textual criticism it will have an objective form), and then move on to dogmatics. Osborne[21] goes on usefully to show how these iterations play out through five separate disciplines: exegesis, Biblical theology, historical theology, systematic theology and practical theology.

* * *

When correctly applied as tools for checking and explicating deductive chains of inference, mathematics and logic are as far as we know infallible. Torrance may or may not be correct in saying that different

---

[20] Bultmann concurs, 1984, p. 63.
[21] Osborne, 2006, p. 351.

systems of logic simply *develop* that of Aristotle rather than *falsify* it[22] – multi-valued logics would seem to call for a more radical realignment. What is true is that such conversations have stayed mainly within the circle of logicians, and have not substantively revised the theories of natural scientists. The development of quantum logic is perhaps the one case where that question remains an open one, and incapable of being settled without a corresponding consensus concerning the interpretation of quantum theory.

What is therefore of greater importance for the natural scientist are the circumstances and protocols whereby logic and mathematical signifiers might correspond to physical realities,[23] which as in the case of non-Euclidean geometry's application within general relativity theory, are so resistant to empirical or theoretic examination, that problems may slip through unnoticed, literally for millennia. Non-Euclidean geometry does not logically contradict Euclidean geometry, for each are built on divergent axiomatics, it is the applicability to physical situations which has been called into question – and that question can only be resolved hermeneutically.

Torrance, taking up Wigner's 'unreasonable effectiveness'[24] of mathematics, fords the ditch with the doctrine of creation, in which by an act of grace, number has somehow become embedded in the material existence of the universe.[25]

> Number is the form of rationality of the inherent order which the material universe displays in so far as it is quantifiable, subject to mathematical analysis and computation. The more we immerse ourselves in physical science the more we are inevitably engaged with number.[26]

This is a strictly Biblical notion of creation, which maintains that whilst it may be infected with particular and structural sinfulness, even so the initial goodness of God in ordering creation and gifting it with its own rational order can still, with theological discipline, be discerned and relied upon, because it is the act of a God who is always the faithful party in his covenantal commitments.

In terms of mathematical quantifiers however, in devotion to its object, theology must reckon with the fact that the only mathematical

[22] Torrance, 1969, p. 271.
[23] Duhem, 1954, p. 108.
[24] See for instance McGrath, 1999, p. 225.
[25] Torrance, 1969, p. 94.
[26] Torrance, 1985, p. 194.

sign which in any way approaches that object is ∞, the quantity which mathematics finds so singularly difficult to handle.

Thus though mathematics and logic seem somehow (we must agree with Torrance following Wigner that the question of exactly how has only extra-logical answers) to relate to reality in such a way that once that applicability is established, their inferential chains are certain, this is only so because mathematics and logic have been abstracted to the extent that nothing within their symbols and variables has any necessary intrinsic empirical reference. Thus the truths of mathematics are often called analytic or tautologous,[27] and rightly so, in the sense that their deductive results are achieved, without at any point employing inductively comprehended features of the physical world -

> it must be pointed out that purely abstract symbolic statements of this kind state nothing about the world of reality and therefore mean nothing, even though they have an important role to play as technical instruments for testing chains of inference.[28]

However mathematics and logic are the *only* sciences possessing this property,[29] and so their rules of applicability can only be established within the synthetic world of empirical experience. Polkinghorne's[30] insistence upon an isomorphic relation between mathematics and the physical sciences is too strong. Given that, as Polkinghorne rightly says, mathematical forms "reflect and point to ontic structures in nature," the question remains *how* they reflect and point, to which the answer is not always self-evident and therefore the relation not isomorphic. Furthermore he goes on to say that these are also determined by the "mode of cognition" of the observer (which is partly socially determined), and therefore leaves the door ajar to the possibility of a degree of flexibility in viable schemas of applicability, a flexibility not available in an isomorphism, unless we are limiting ourselves to a mathematically trivial degree of flexibility. The question of cogency in quantification will therefore contain an element of hermeneutical ambiguity, which must rely on the intuition of the scientist.[31]

Thus no sciences are wholly devoid of the implications of logic and mathematics, yet equally no sciences are devoid of more hermeneutical

---

[27] Ritchie, 1948, p. 71ff.
[28] Torrance, 1969, p. 225.
[29] Duhem, 1954, p. 266.
[30] Polkinghorne, 1991, p. 39f.
[31] Kuhn, *The Essential Tension*, p. xiii refers explicitly to the 'hermeneutical method'.

questions of applicability, and thus cannot be reduced to logic or mathematics. However there will also be a radical divergence in the efficacy and propriety of logical and mathematical techniques within each science, and faithfulness to object will determine where each science sits on this spectrum. Therefore the possibility of a dichotomy between mathematical and hermeneutical disciplines solving the demarcation question of science fails on both sides: for however much the proportions vary, every science has hermeneutical, intuitive, inductive, logical and deductive components.

Whereas many philosophers of science, for instance the logical positivists and Popper, were greatly concerned with such questions of demarcation,[32] Feyerabend is clear not only that such a demarcation does not exist, (and that where *de facto* it does humanity would be better rid of it), but also that there is nothing so different between the science and the arts,[33] and indeed science and fairy tales[34] and myths.[35] This has implications for how society assesses the epistemological value of natural science: Feyerabend's aim is to cut it down to size, claiming that science has become too powerful.[36]

The focus of our discussion of hermeneutics is therefore on the relationship between interpretation and science, and the question of demarcation implicit therein. It is as well therefore to set out what function and results we need from our discussion of hermeneutics, so as to give it reasonable boundaries.

- Uncontroversially, we have asserted that hermeneutics is a modality of cognition found deeply embedded in theological discourse.[37]
- Unlike logic and mathematics, hermeneutics cannot function in a merely deductive manner, from premiss to conclusion. A variety of 'hermeneutical circles' exist.
- Given that no natural science can be constructed solely in terms of mathematical or logical symbolism, every science has a hermeneutical component – though the importance of that component will vary widely.

---

[32] Popper, 1992, p. 311.
[33] Feyerabend, FTR, p. 294 also Polyani & Prosch, 1977, p. 65.
[34] Feyerabend, AM, p. 298, also p. 52.
[35] Feyerabend, KSR, p. 60.
[36] Feyerabend, AM, p. 217.
[37] "The hermeneutical problem is one of the basic primal problems of Christian theology in general", Blank, in Küng and Tracy (eds.), 1989, p. 262.

Feyerabend's assertion that scientists are simply creating works of art[38] is one extreme of this spectrum. Torrance tries to keep together the unity and the diversity of the sciences: the unity of knowledge under the gracious act of creation, and the diversity of the sciences by reference to the logics of number, word, organism, persons, grace and beauty.[39]

- We will therefore agree with Gadamer, contra Dilthey, who insists that this hermeneutical component exists in all attempts at human understanding, rather than marking a possible demarcation point between the arts and the sciences, or even between the textual and mathematical sciences.

> Observer and object are linked in a mutual relationship. The nature of the object controls what can be known about it and the way in which knowledge must be expressed ... a kind of version of the hermeneutical circle is involved.[40]

* * *

Feyerabend's contention that natural scientists too produce works of art is a deconstructive extrapolation, somewhat reckless, from his earlier notion of the essential unity of the sciences (which he is, as ever ironically, later a fierce critic of[41]), a unity which coincides with Torrance's unitary theory of knowledge, and indeed with Gadamer's universal hermeneutics. In the excellent volume *Hermeneutics versus Science?* to which Gadamer contributed, the last words contain signature themes of Feyerabend (in fact they are from Stegmüller):

> Further, we say that certain forms of the circle of understanding are closely connected to the problem of the theory-ladenness of observation.
>
> And finally we have come to see that no form of this circle at all can be used systematically to demarcate, much less 'to indicate the special feature' of the humanities or historical sciences vis-à-vis the natural or non historical sciences. For to stress this point once more, every branch of science is potentially threatened by all these difficulties.[42]

---

38 Feyerabend, FTR p. 294.
39 Torrance, 1969, p. 205ff.
40 Polkinghorne, 1991, p. 15, also p. 6.
41 Preston, 1997, p. 180.
42 Stegmüller in Connolly & Keutner (eds.), 1988, p. 152.

Gadamer, contrary to those such as Dilthey who espoused the opposite view, maintained that hermeneutics was not simply a literary technique, but a fundamental philosophical condition of all acts of understanding.

> Understanding, says Gadamer, is always an historical, dialectical, linguistic event – in the sciences, in the humanities, in the kitchen. Hermeneutics is the ontology and phenomenology of understanding.[43]

Thus the 'method' of Gadamer's *Truth and Method* is an ironic one, for the hermeneutical circle is not method but condition.[44]

Whilst some philosophers and theologians (Heidegger, Bultmann, Foucault), especially those with a background in textual disciplines, often tend to possess a somewhat naïvely exalted reverence for the findings of natural science, Feyerabend redresses the balance, essentially by extending Freud's, Marx's and Nietzsche's hermeneutic of suspicion to scientific texts. Whereas those others might mark a strong distinction between the 'non-dubious' findings of natural scientists, and the hermeneutical sciences and creative arts, Feyerabend maintains that even within natural science under the veneer of the 'objective' meaning of the text is a sub-text of equal or greater significance. The 'creations' of natural science are not simply statements and theories about natural objects and phenomena, they too implicitly harbour the political presuppositions of the scientific and cultural 'tradition' which produced them.

The question remains, however, as to the balance of powers between text and posited sub-text. Freud, for example, is dealing with cases where the text is pathological to the extent of aphonia, so that the literal or primary text must be disregarded in order to uncover the 'true' (sub-) text. If, to take Feyerabend's example of Chinese and Western medicine,[45] Western medicine's account of itself as a means simply for engendering 'health' (itself an unarguably tradition bound notion) must not be taken at face value, but as cover for its function as a vehicle of capitalistic imperialism, the question then becomes: is that condition 'pathological'? Whilst Derrida following Nietzsche posits that most dialogues are not decided by rational agreement, but by non rational forces and drives,[46] Feyerabend, particularly in the paucity of

[43] Palmer, 1969, p. 215.
[44] Palmer, 1969, p. 178.
[45] Feyerabend, AM, p. 36f.
[46] Schmidt, 2006, p. 166.

his examples, fails to secure that conclusion for natural science. The problem is that Feyerabend does not distinguish, as Mannheim for instance does,[47] between 'unmasking' and 'refuting' a thesis. Feyerabend rather assumes that once 'unmasked' – for instance that observations are theory-infected or that social and political factors have played a part in scientific research – then any truth value within those theories has been fundamentally devalued, rather than evaluate precisely the extent of that devaluation and its consequences. After all once diagnosed some theory-infections can be easily treated. Anarchy, if it is 'medicine' as Feyerabend maintains, then it is pretty strong stuff, and like chemotherapy the damage may sometimes outweigh the benefit in all but the most serious pathology – a clinical decision is called for and Feyerabend's is unwarrantedly drastic.

The task of determining this relation between the text and sub-text of natural science is therefore one with hermeneutical components. That is not to pre-judge the outcome of the Gadamer / Dilthey debate, but to note that if we play along with Feyerabend then we enter Gadamer's territory. The question becomes: can we circumnavigate the sub-textual lesion – surgically remove it – so that the priority and integrity of the text is restored (which may require an ongoing programme of hermeneutical therapy), or is the sub-text so deeply imbedded that in fact a pathological inversion has occurred, so that any attempt to kill off the text's sub-textual baggage will also kill off its ability to maintain objective reference, the life-blood of scientific texts?

Torrance would certainly agree with Feyerabend that natural science has a sub-text – in particular the metaphysical 'climate' described earlier which is implicitly imbedded within scientific conversation – making some considerable leaps of imagination in that regard: he proposes that John Philoponus' metaphysical critique of Aristotle enabled him to have a premonitory insight of relativity theory, and that Newton's mechanistic view of time and space was rooted in an Arian Christology (see 'Fluid axioms' in chapter eight). However his hermeneutics of suspicion is a far more moderate affair than that of Feyerabend.

To the casual observer it may appear obvious that any cultural artefacts embedded within the flow of Western natural science, have, even if real and present, been of negligible significance, compared with the fundamental advances in human understanding which that science has provided. The examples for and against are self-evident: space

[47] In Hacking, 2000, p. 56.

travel, the internet, consumer electronics, technological medicine, increased agricultural yields, vs industrial weapons of destruction, ecological damage, collateral damage to fragile systems of tacit and intuitive understanding, and more importantly the economic and political systems which have fostered and accompanied this particular set of Western 'sciences'. Our task is not to weigh these two sets of opposing examples in the scales (whose scales?) of human progress, for that is with Feyerabend, to confuse the ethical with the epistemological – though we note that increased awareness of the damage caused by the dominance of the technological paradigm has been part of the transition from modernism to post-modernism.

Epistemically the relation of text and sub-text within natural science, in common with any such hermeneutical question, can only be understood by analysis and immersion in the text and its context (the anthropological method of Evans-Pritchard, which is so important to Feyerabend) – only by *doing* theology and natural science, can we understand the proper relationship between them. Only by immersing ourselves in that on-going actual engagement can we know whether our thesis is correct or incorrect. It is only with Freud's problem-solving interpretation of Hamlet, and similar 'successes' in Nietzsche and Marx, that the hermeneutics of suspicion are given credence within the interpretative community. So does Feyerabend's analysis of natural scientific discoveries justify a hermeneutic of suspicion this potent?

The general verdict within the philosophy of science community is that Feyerabend's analysis of particular scientific discoveries is neither deep nor wide, and though others such as Kuhn, Polanyi, Putnam have been sympathetic to some of his ideas: incommensurability, lack of universal method, democratic control, theory-ladenness of observation – few have judged his arguments for the radical versions of these theses to be convincing. They have prescribed caution rather than anarchy. Incommensurability has been interpreted in a variety of ways; lack of universal method does not threaten critical rationalists who insist on a few guiding principles, combined with methods which are immanent to the various sciences (as Torrance); democratic control is an ethical and political issue, and Feyerabend's conflation of it with an epistemological justification of voluntarism is simply regarded by most (including me) as a category error; theory-ladenness does not entail the more vicious circularity of theory-determinedness.

Feyerabend's failure is in large part due to his habit of using one, often rather dubious, historical example[48] for making a more general case.[49] Feyerabend's critique of rationality far too often consists of showing that natural scientists do not always follow the rules. That defeats only the straw man of rationalism – the real man of rationalism demands only that it would generally be better overall if natural scientists acted methodically.[50] That real man cannot be defeated by a counter-example or two. In fact the irony of the situation is that Feyerabend tends to proceed in a falsificationist manner when it comes to his historical counter-examples – whilst Feyerabend (along with many other philosophers of science) rejects Popper's basic thesis that it is possible for one counter-example to be a decisive refutation of a natural scientific theory, that is exactly the kind of schema Feyerabend employs when analysing the history of science. The most crucial example of this is that of Galileo, which disproportionately dominates *Against Method*. The dilemma for Feyerabend is that these, at the time reactionary, views are now common currency in Galileo scholarship. However this absorption is a pyrrhic victory for Feyerabend, because though they might mark a vindication of his interpretation of Galileo, the fact that his more radical manifesto has been largely ignored leaves him out of ammunition. Rationalists might have become slightly more cautious and critical, but they have not become anarchists. Galileo is not to be Feyerabend's Hamlet, in scientific terms it is anomaly rather than falsifier.

However the consolation for Feyerabend is that whilst he may not have made his case for anarchism in this particular case, he has forced us to submit the text of natural science to a hermeneutic of suspicion, even if the conclusion is that no charges are to be pressed. In other words more naïve versions of rationalism – those that hoped for a prior explication of scientific method, an *a priori* demarcation criterion, and a non-hermeneutical account of its contents – have been disappointed. It is only by a critical rationalist engagement with natural science itself that we have been able to exonerate it. Is this not a hermeneutical circle of just the sort which rationalist defenders of science wish to avoid? The answer to this weaker assertion must be a provisional 'yes' – the

[48] Lauden, 1996, p. 102.
[49] Feyerabend, AM, p. 27.
[50] Lauden, 1996, p. 103.

achievements and results of natural science do not entirely escape the hermeneutical circle which Dilthey hoped that they might. The situation is therefore far more Gadamer-like.

* * *

It is also important to understand what are the epistemic consequences of introducing the term 'hermeneutical circle' into a discussion about scientific method. With Gadamer, both Feyerabend and Torrance are contending that the hermeneutical circle as traditionally defined is simply one paradigmatic example of a more general epistemological condition, which pervades all human understanding, and occurs in many and varied guises. Stegmüller is an exponent of this view, offering us six examples.[51] Polkinghorne also says:

> No quest for truth can escape from the necessity of this hermeneutical circle, linking the encounter with reality to an interpretative point of view, so that they are joined in a relationship of mutual illumination and correction.[52]

'Circular' when applied to logical arguments is generally and correctly understood pejoratively. Deductively it is the inclusion within one's axioms, whether implicitly or explicitly, of that which one is trying to prove; inductively it is the inclusion within an observation, implicit or explicit, of that which belongs to the theory one wishes to verify. As we have already seen Feyerabend is keen to accuse natural science of the latter offence, particularly natural science as it is construed by logical positivism. Such circularities are properly styled 'vicious' and logical arguments which employ them are invalid – though of course true conclusions may arise from invalid arguments. However the sorts of circles of understanding to which we are referring are not necessarily of these kinds, and indeed the point of the hermeneutical circle is that it calls into question the syllogistic premisses–argument–conclusion structure of the knowledge – if we admit Torrance his fluid axioms (see chapter eight) and his extra-logicality then we draw the sting of the charge of logical invalidity, as long as the appropriateness of the scientific predicate has not been thus compromised. The circle of understanding is only vicious if we specify in advance that we are using certain forms of argument.[53]

[51] Stegmüller in J. Connolly & T. Keutner (eds.), 1988, p. 110.
[52] Polkinghorne, 2007, p. 8.
[53] Habermas, 1972, p. 171ff; Pannenberg, 1976, p. 200.

Thus I am not seeking to offer either a thorough history of hermeneutics or to recommend a set of hermeneutical principles or techniques. What I am trying to show, much in the style of Gadamer's philosophical hermeneutics, is that hermeneutical circles are embedded in all areas of human understanding – certainly the natural scientific and the theological – and that whilst they deny the possibility of reduction to the linear rationalist progression of classical foundationalism, they are nonetheless compatible with, indeed a necessary part of, the sort of realist anti-/fideistic-foundationalism which Torrance advocates. Hermeneutics therefore, for Feyerabend and Torrance, is both a common component across all sciences, but also in the hermeneutical relation of object and method, object and rationality, it provides an account of the differences of method and rationality emergent from within that dialectic. In our final conclusions we will seek to explain why it is that this unity-in-diversity plays out in such different ways in Torrance and Feyerabend; and how Torrance can justify his particular path by a threefold appeal to an underlying onto-theological basis, which in turn unifies its diversity of method as correlated to a corresponding diversity of subject matter, and finally an epistemological disposition motivated by and predicated upon his theo-ethical imperatives. Our final contention will be that the coherence of such a three-fold structure is enough to defend Torrance's theological epistemology from Feyerabend's anarchism.

* * *

Feyerabend is, of course, not generally interested in Biblical hermeneutics, but nonetheless his exercising of a 'hermeneutics of suspicion' within the field of natural science means that we must still make, where possible, suitable connections. As we prepare to turn to Torrance's hermeneutics, we must therefore prepare the ground by noting three explicit points Feyerabend makes with regard to Biblical hermeneutics.

Firstly, Feyerabend praises theology[54] for at least getting its methodology out in the open, in a way in which natural science often fails to do. Secondly, he is no friend of demythologizers, such as Bultmann, the one theologian (as far as I have noted) whom Feyerabend mentions more than once.[55] Thirdly, part of Feyerabend's fire is directed at those

---

[54] Feyerabend, FTR, p. 251 – hermeneutics as a discipline grows out of theology.
[55] Feyerabend, FTR, p. 260; Feyerabend, KSR, p. 143.

who too lazily assert the superiority of subjects studied within the academy. If traditions from beyond the academy are to be taken seriously as human attempts at understanding the world, then the corollary is that religious movements from beyond the academy should also be taken seriously as attempts to understand God.[56] So there can be no doubt that his praise for liberation theology[57] is genuine, and that its chief recommendation is its elevation of praxis over theory, mirrored in Feyerabend's own overarching pragmatism, an elevation rejected by Torrance.

Torrance is primarily a dogmatic theologian rather than a Biblical interpreter, but it is nonetheless of fundamental importance to him that these two disciplines cannot be separated:

> Biblical interpretations cannot be carried out faithfully unless exegetical and theological work go hand in hand, if only because it is impossible to engage in exegesis without epistemological assumptions which require critical testing in hard theological thinking, but also because theological statements must be contained within the limits of the divine revelation mediated to us through the Scriptures.[58]

This statement appears at the end of his exploration of patristic hermeneutics (*Divine Meaning)* which alongside *The Hermeneutics of John Calvin* (1988) are the two volumes explicitly dedicated to the hermeneutical field, though similar concerns are present in all his work, in addition to his role as an ordained minister of the Word throughout his adult life.

Where does Torrance fit in with traditional schools of hermeneutics? We have already implied that he shares with Feyerabend a belief in universal hermeneutics.[59] What about matters regarding the authority of author, text and reader?[60] These questions must be framed within his two overriding conditions of interpretation: the Biblical material as a whole must be understood as a self-authenticating and self-interpreting world[61]; and secondly that same material witnesses to the revelation

[56] Feyerabend, FTR, p. 260 – Feyerabend's complaint about the timidity of academic natural scientists' theological counterparts, is also found in Heidegger – see Geldhof, 2007, p. 52.

[57] Feyerabend, AM, p. 248; It is of course a separate question as to exactly how this 'grass roots' movement relates to its defenders within the academy.

[58] Torrance, 1995, p. 426.

[59] Torrance, 1969, p. 316.

[60] Brett in Watson (ed.), 1993, p. 14.

[61] "Verification is the *autousia* and *autoexousia* of God through Scripture" – Colyer, 2001a, p. 370.

of the Truth of God in Jesus Christ, and is intra-textually assessed according to this plumb-line. Ultimately potency of reference determines correct hermeneutics. Authorial intention, the discovery of which will of course generally aid understanding, can also function as a diversion from the witnessing power of the text, so that even the person of the evangelist may stand in the way of proper understanding of the text:

> We use language scientifically only when the primary intention is brought into play and its deliberate reference is taken seriously. We cannot eliminate the fact that St Mark put something of himself into his witness and into his writing, yet his primary intention was to speak not of himself but of Jesus Christ. If we are to deal faithfully with St Mark, we must look at the reality to which he points and interpret what he has to say primarily in the light of it – otherwise we fall down badly in regard to basic scientific procedure.[62]

The hermeneutic principle of interpreting the text of the Bible from within the world of the Bible, Torrance calls 'indwelling',[63] resonant of the anthropological method of Evans-Pritchard which serves such an important role in Feyerabend's 'methodology' – *complete with analogous shortcomings*. Like Barth, Torrance discovers in his reading of the Bible a 'strange new world'[64] and thenceforward it is fundamentally our horizon which must be moulded by the 'conceptual foundationalism'[65] found therein. Whilst the horizon of our pre-understandings will always frame how we read a text, the Bible must be allowed to speak with its own voice and on its own terms. Even if our present proclamation of the gospel message is retranslated into the conceptualities and language of our horizon, so it is *these* which are called into question by the Biblical narrative, just as new wine demands new wine skins.[66] The task of the scientific theologian is therefore to indwell this 'strange new world', constantly circuiting the hermeneutical spiral of whole and part:

> In studying the Scriptures we have to pay attention not only to what is written in them but to the acts of God to which they bear witness – i.e. to the whole economy of salvation in the birth, life, death, resurrection and

[62] Torrance, 1971, p. 38.
[63] Torrance, 1996, p. 37.
[64] In Barth, 1928, p. 28; Torrance, 1971, p. 19.
[65] MacDonald, 2000, p. 115.
[66] For Torrance's use of this Gospel metaphor, see e.g. 1969, p. 25, 280.

> ascension of the Incarnate Son of God. Interpretation is therefore inevitably theological and takes place properly within the scope of faith and its basic doctrines.[67]

Yet also,

> we must recall at this point the basic rule of all Church dogmatics: that no single item of Christian doctrine is legitimately grounded or rightly developed or expounded, unless it can of itself be understood and explained as a part of the responsibility laid upon the hearing and teaching Church towards the self-revelation of God attested in Holy Scripture.[68]

Torrance sums up this approach as a 'theological and unitary interpretation,'[69] in which whilst we might avail ourselves of the tools of any school of interpretative analysis, those tools must only and self-consciously be used as such,[70] and not as means whereby distortive philosophical assumptions are smuggled into the text – tools are sometimes quickly made blunt if used on materials for which they were not designed and have to be hastily discarded. Thus the Biblical theologian is set free to use whatever tools he finds suitable during his work, as long as he remembers that they are just tools and not instructions.

In his work on the hermeneutics of John Calvin, Torrance proposes that once we undertake this interpretative approach, we will discover that it is in its reference to Jesus Christ that we find the christological centre of and key (*scopus*) to Biblical interpretation:

> Christ is himself the scope of the sacred writings, the mark at which they point, and therefore provides the point of reference which gives harmony or coherence to all the books of the Old and New Testaments alike.[71]

Jeanrond[72] similarly posits a distinction between micro- (the interpretation of particular texts) and macro- ('the interpretation of the universe') hermeneutics, and furthermore insists that theology must concern itself with the latter, but only by working through the former. He terms the latter task 'hermeneutical theology', being the immediate method by which we tackle the former wider task of 'theological

---

67 Torrance, 1995, p. 273, also Osborne, 2006, p. 139.
68 Barth, CD II/2, p. 35.
69 Torrance, 1996, p. 48.
70 MacKinnon in Fergusson (ed.), 2007, p. 209.
71 Torrance, 1988, p. 51.
72 Jeanrond, 1994, p. 8.

hermeneutics', wherein for the Christian, Christ emerges as the *scopus* by which the universe is at least implicitly interpreted.

According to Torrance the chief threat to such Christocentricism[73] is our self-centredness and self-fascination: "in all knowing and interpreting he (i.e. the Biblical theologian) must work from a centre in Christ and not in himself." Thus for instance, the New Testament's appropriation of Hebrew Scripture texts as referent to Jesus Christ, thereby subverting the notion of author intentionality – noting that the whole notion of 'authorship' of the prophetic material is *itself the* hermeneutical question – is for Torrance entirely legitimate.

If Jesus Christ is to be the chief interpretative key for the Biblical material, it might therefore be assumed that the 'verification' of Jesus Christ on extra-Biblical grounds, as attempted by Pannenberg, might be of concern to Torrance. However Barth and Torrance's insistence on the foundational nature of the Biblical texts is such that it implies not only that those texts are self-interpreting, but that they are also self-authenticating. To prove the existence of God, or the truth of the message of Jesus Christ on other grounds, can only be a subverting of the

[73] Torrance 1969, p. 81 – the unwarranted concentration on the 'for me' of 'Christ loved me and gave himself for me' is seen by Torrance as being behind the whole liberal protestant tradition which exchanged the gains of Calvin and Luther for the humanist individualism of the Renaissance – more present in Lutheranism generally than Calvinism. In the same way the Roman Church had assumed the rights of *veritas Dei* (Torrance, 1988, p. 64), so had 'modern man' in liberal Protestantism.

Thiselton, 1992, p. 531:

> "Very often in religious groups an individual is encouraged to "frame" the Biblical text with reference to the narrative history of personal testimony, and to "read" the text as "what the text *means to me*". If this is undertaken within a frame of corporate evaluation and testing, the life-experience in question may enhance pre-understanding and weave meaning and textual force with emotional warmth and practices in life. But without any principle of *suspicion*, in Gadamer's terminology a *premature* fusion of horizons will take place *before* readers have listened in openness with respect for the tension between the horizons of the text and the horizon of the reader. The textual horizon has collapsed into that of the reader's narrative biography, and is unable to do more than to reflect back his of her own values and desires.
>
> This example exposes the ultimately *socio-pragmatic* status of Hollands' theoretical model. In the end, if "we use the literary work … to *replicate ourselves,*" as in Fish and Rorty, the text can never transform us and correct us "*from outside*." There can be no prophetic address "from beyond." This may still leave room for a measure of *creativity and surprise* in *literary* reading. For *in such cases it does not profoundly matter whether it is ultimately the self* who brings about its own creative discoveries. But in the case of many Biblical texts, theological truth-claims

Truth.[74] However that means that any defence against Pannenberg's charge of 'illusion and caprice'[75] has also been severely weakened.

Barth and Torrance's abhorrence of natural theology (less strident in the latter) implies that the truth of God's revelation is only known because it is God's revelation, and the truth of his being is only known in his being as Truth, hence the reluctance to justify these things on other grounds.

The 'strange new world within the Bible', constitutive of a conceptual foundationalism, which we cannot circumnavigate but must submit to, might therefore be taken to indicate a high regard for the authority of the text. If however intra-textual coherence is disjunctively given priority over extra-textual reference, then Torrance would deny the text that degree of autonomy – the text serves as witness to that to which it refers and corresponds. From there it *derives* its authority, on a realist basis, not in its relations with other texts either internal or external to the Biblical material.

So in 'literal' terms 'reader response' best describes Torrance's position, but as the polar opposite to how that term is more commonly understood.[76] For Torrance the text as it witnesses to the Word of God

---

> constitute more than triggers to set self-discovery in motion (even if they are not less than this). If such concepts as "grace" or "revelation" have any currency, texts of this kind speak not *from the self*, but *from beyond the self*."

[74] Barth, 1949, p. 36–37.

[75] Holwerda in Plantinga & Wolterstorff (eds.), 1983, p. 285; see also Pannenberg, 1968, p. 109.

[76] Though Thiselton is generally scrupulously fair in his survey of hermeneutical traditions, one cannot but feel that with reader response his patience begins to wear rather thin:

> The thrust of his thesis [Fish] "going down the Anti-Formalist Road" (1989) entirely misses the points made by Wittgenstein about concepts with blurred edges, family resemblances, and the common behaviour of humankind. Wittgenstein in effect *rejects* Fish's artificial alternative that *either* we have a formal system (like the *Tractatus*) *or* there are no inter-penetrating or overlapping regularities of a stable nature which transcend the boundaries of a single language-game or context-relative social community. The supposed alternative of *either* formal system *or* social relativism without trans-contextual critique is a false one. (Thiselton, 1998, p. 516 – his italics – his criticism of Fish's misuse of Wittgenstein could equally well be applied to Feyerabend).

The weaker form (see Osborne, 2006, p. 479 – distinction here between Iser (weak) and Fish (strong)) of reader-response theory appears to say mildly interesting things, which hermeneutical scholars have understood for decades: people read and understand different texts in different ways, and this variety is often a result of the different pre-understandings and life-worlds which they bring to the text. Gadamer's image of the two horizons which (at the appropriate point – there can be a premature

stands in judgement over its reader, demanding submission and repentance, and like Job, the reader finds that ultimately his questions are turned back upon himself. It invites the reader to enter its world, to be converted to *its* way of thinking, indeed to its rationality, and the 'logic

merging) merge to create a fused understanding of the text, seems to point roughly in this direction. But when we move beyond this, to saying that the reader somehow creates or co-creates the meaning of the text, reader-response goes too far. Although a presuppositionless reading of a text is impossible, and bearing in mind the known and unknown intertextual relationships for both author and interpreter which shape understanding of the text, it is nonetheless correct exegetical procedure to make explicit our pre-understanding, in order to neutralize its distorting properties. If we come to a text, for instance, with a 'container' view of space – a view not shared by the text (say a text which posits a relativistic space-time continuum), and then having read the text, leave with our view unchallenged, in what sense can we be said to have 'responded' to the text? As Ricoeur points out, the danger of reader-response theories is that rather than 'responding to the text', they will in fact simply project their idolatrous desires onto it:

> But if this is the case, the reading of Biblical text, as Paul Ricoeur so cogently argues *can result in idolatry. We can project our own interests, desires, and selfhood onto that which the Biblical text proclaims.* We can thereby unwittingly re-create and *"construct" God in our own image through our reading processes.* (Thiselton, 1992, p. 530, his italics).

A response implies that our understanding is changed in a way that it would not have been if we had not read the text. It implies that the text has somehow acted upon us, has changed us. How can this occur if we deny the text any meaning external to us? The implication of the notion of 'response' is not that we shall be over implicated in the creation of textual-meanings, but rather that we shall return to the rather naïve contrary view that texts have fixed meanings which they impose upon us.

Reader response hermeneutics is often associated with particular theologies. However the stronger version of this style of hermeneutics would actually undermine such theologies. It is one thing to argue that the particular experiences of oppression and alienation suffered by for instance women and black people, are illuminating to the reading of the Biblical text, and may help us to understand how the text has been misunderstood and misused in the past. However the force of this valuable insight is lost unless the dynamic of oppression and liberation is understood as something found objectively in the text. Otherwise the oppressor is just as legitimately able to read the text as oppressor, and (co)-create a meaning adequate to his purposes and concerns. Contexts are particular, but the interpretation of Biblical texts is only ever mediated by relating those particulars to the multi-contextual tradition of the Christian church (Rae, 2005, p. 138–9).

How has reader-response criticism at its most radical led us to this dead end? I think the inter-related reasons are four fold: an overdeveloped analogy, a simple piece of mathematical misunderstanding, a naturalistic fallacy of relativism, and a misunderstanding of Wittgenstein. There is an element of truth in each of these which is lost as the strong reader response theory gets carried away.

Firstly the reading of a text is frequently compared to the performance of a piece of music. Music is not, so the analogy goes, simply dots upon a page, it is only music when it is performed – and each performance will be slightly different, with each interpreter playing it slightly differently (Thiselton, 1992, p. 99, 587). The main

of grace': "Hence the demand of the Gospel for repentance on our part, for radical change even in the inner slant of our minds and in the structural capacities of our reason."[77] "Reading Scripture aright is not finally a matter of technique but of grace."[78]

Though critical of the crude scientism of demythologising[79] which undermined Bultmann's hermeneutics, Torrance is largely in agreement with his existential interpretation deriving from Kierkegaard, and which understands the Biblical message as proclamation calling for a decision.[80] That decision is a far more active response than simple acquiescence to necessity,[81] being rather a positive *response* to a call from beyond us, and not something internal to the dynamics of our consciousness: it is Clement's 'free assent of the mind' and Duns Scotus'

---

problem is that the analogy does not correspond with the stronger version of reader-response. Firstly the score does constrain the performer to such an extent that every piece is recognisable from its score. We can identify bad performances, wrong notes, and people playing the wrong piece of music by mistake. The variations of possible performance are infinite and each time unique yet tightly bounded by the objectivity of the score. Musicologists *can and more often than not do* compare the operas of Monterverdi and Gluck, without having to specify which particular recordings and performances they are referring to.

This leads to the second confusion, the elementary mathematical confusion of the infinite and the unbounded. However small we choose to make the interval, there are always an infinite number of real numbers between any two finite numbers. Reader-response theories point out that every readers interpretation will be different (whether every reader will be able to express or tell the difference is another matter), which is a perfectly valid point to make, but that fact has absolutely no implications for the range of correct interpretations, which may in fact be very narrow indeed (Osborne, 2006, p. 487–8).

Thus our third point is that reader-response theories often consist of an argument from variety so beloved of Feyerabend and his various relativist allies – there are many different ways of interpreting the text implies that there are many ways in which the text should properly be interpreted. That is a fallacy.

Fourthly as Thiselton points out, these misunderstandings often emerge from an improper Feyerabendian extension of the later Wittgenstein's ideas, an extension that those very writings deny. The blurred edges of language games, is not to deny that once everybody concerned is agreed which game we are playing, we can then get on with playing it, and that it is then possible to call 'foul' and argue that a mistake has been made. The possibility of *mistaken* interpretation by a reader, seems to me to be litmus test of the strength of reader-response theories – if mistaken readings are possible, then the theories don't say very much which is original. Conversely if they deny the possibility of mistaken readings then such theories are wrong.

[77] Torrance, 1969, p. 49; Torrance, 1969, p. 253.

[78] Rae, 2005 p. 145.

[79] Though distinct from rationalising (Bultmann, 1984, p. 104), it is still infected with the positivistic scientism of its day.

[80] Ricoeur, 1981, p. 69.

[81] Torrance, 1969, p. 214.

'voluntary knowing.'[82] However this must not be mistaken for a projection of our own concerns and conceptions onto the text. Existential hermeneutics, with a fork in the road at Kierkegaard, is divided, for Torrance, between those who recognize the existential element in hermeneutics and those who see hermeneutics simply as existential, for as Thiselton remarks: "The one sidedness of existentialist interpretations would be disastrous for Christian theology and do violence to our understanding of the New Testament if it is regarded as a comprehensive hermeneutic."[83]

The problem with positioning Torrance's hermeneutics in traditional categories is that his scientific realism is so dominant, that matters of interpretation are subsumed by matter of reference, so that the interpretative question is always 'what is the truth concerning that which this text witnesses to, regardless of the foibles, accretions and pre-judices of the author, the text itself, and particularly the reader?'[84] Any principle or authority, however noble it may appear – the evangelist, the sacred text, our own concerns – which stands over and against our proper cognisance of that object is ultimately an idol to be dispensed with – remembering that it is the imperative of faithfulness to object which is constitutive of sound scientific method for Torrance. If that imbeds us uncomfortably deeply into the hermeneutical circle – fidelity of reference is the interpretative key of a text to whose 'bondage' (Barth[85]) we are already subject to for knowledge of that object – then so be it. That is not to say that we cannot find other meanings and reference within the Biblical text, and indeed they may be illuminative in a peripheral sort of way, but if we stare too fixedly at them we are liable to become blinded when we again turn to face the main source of illumination:

> In no way can the light of the Scriptures substitute for the light of Christ, for they are entirely subordinate to his light and are themselves light only as they are lit by his light. Indeed it may be said that if the Scriptures are treated as having a light inherent in themselves, they are deprived of their true light which they have by reflecting the Light of Christ beyond

[82] Torrance, 1984, p. 197.
[83] Thiselton, 1992, p. 282.
[84] If this approach seems somewhat naïve with regard to later linguistic philosophy, we are tempted by Moore's riposte concerning Paul: "If Paul mistakenly thought that 'discerning the body' involved moving from signifier to signified, too bad he hadn't read Derrida," 2003, p. 112.
[85] Barth, CD I/2, p. 208.

> themselves – and then the light that is in them is turned into a kind of darkness.[86]

Like our subjectivity, the Scriptures possess a derivative authority, being only trustworthy in so far that such authority flows dependently from the ultimate authority which is God's alone.[87] Our knowledge of God will be mediate yet realist, because we shall always be trying to describe a reality which lies somewhere at the edge of our vision, and yet we must never take our eyes off that limit point, and seek to describe a more accessible target on its own terms.[88] We examine the craters, but only for clues as to the nature of the initial explosion, not because we are interested in craters in and of themselves.

Torrance's hermeneutics is therefore imbedded within his particular Barthian 'abductive' Scriptural doctrine of revelation, so that the former is challenged if and when the latter is. One might for instance understand revelation as process rather than an event; or one might, often according to denominational preference, order the relation between the different sources of revelation, mediate or otherwise, differently from Torrance – one might talk up the role of the Spirit, religious experience, or the witness of the Church, and play down the importance of the Biblical witness. Such a thorough survey is beyond the scope of this book, as our main concern is to show that the ideas contained within Torrance's own theology cohere with his conception of what makes theology a science – we are not here to defend his doctrine of revelation against all comers, but a note of caution has been sounded.

Neither does the dominance of a referential interpretative key imply that for Torrance the lenses of author, text and reader, are all of equally good or bad quality – indeed far from it. The Biblical text has authority over us and the church, in that it has objectified its role as witness in history, in the way in which we have not. The reader stands *more* under judgement within the triadic relation,[89] because of the obviously distorting effects of wilful and sinful self-projection.

> Hence we must keep a constant check on these refined theological concepts and relations to make sure that they are in definite touch with the ground level of God's actual self-revelation to us, and our evangelical

[86] Torrance, 1982, p. 95.
[87] Torrance, 1982, p. 154.
[88] Barth, CD I/1, p. 268.
[89] Osborne's 'trialogue', Osborne, 2006, p. 516.

> experience of his saving activity in history, and that they remain empirically correlated with the saving truths and events of the Gospels, otherwise they tend to pass over into mythological projections of our own rationalisations into Deity.[90]

This is of course to smuggle a theological anthropology into hermeneutics, the Reformed principle that our depravity extends to our capacity for knowledge,[91] an anthropology which contrasts with Feyerabend's more optimistic one, but it cannot be circumvented, being an integral feature of theological science.

However because it is divinely inspired human witness, the manner in which the Scriptural text becomes *verbum Dei* within the interpretative community lacks the ultimate directness of the prophetic word prefaced: 'Thus says the Lord'. As Ricoeur says:

> In the same way, the relation "human word / word of God" is placed, no longer between the New Testament and the rest of the Bible, no longer even between the New Testament and the rest of culture, but at the very heart of the New Testament. For the believer, the New Testament itself conceals a relation that needs deciphering. This relation is between what can be understood and received as word of God and what is heard as human speaking.[92]

Ricoeur is right in saying that fundamentalist readings of the Bible have mistakenly placed everything indiscriminately within the genre of prophecy – i.e. the 'Thus says the Lord' particular to prophecy becomes prefixed to all the Biblical material – even where a human opinion is contrasted with divine word (e.g. 1 Cor 7:25). "Revelation is the speech of another behind the speech of the prophet."[93] However alongside prophecy the Bible also contains narration, prescription, wisdom and hymnic texts.[94] Certainly we should not overlook the links between 'law and prophets', deny more mediate and ambiguous modes of inspiration, or ignore the injunctions to verify and interpret prophecy – but the point remains: there is something rather self-serving and inauthentic in imagining that a psalm of praise to God ultimately finds its authorship in that God, through a human delegate, rather than a free creative human response to that God – which therefore may contain imperfections.

---

[90] Torrance, 1996, p. 109.
[91] E.g. Calvin, 1960, p. 40.
[92] Ricoeur, 1981, p. 55.
[93] Ricoeur, 1981, p. 75.
[94] Ricoeur, 1981, p. 78.

Thus the interpretative key of genre cannot be known as something prior to the text. Rather there exists a hermeneutical circle by which we can only grasp the genre of a text by reading it, but how we read it will be determined by what genre we think it belongs to. Often this circle dissolves almost immediately, but at other times (e.g. Lewis Carroll, James Joyce's *Ulysses,* Sokal hoax), especially where authors do not strictly observe the usual conventions of genre, real confusion occurs.[95] Anyone reading *Ulysses* in the hope of gaining a basic account of Dublin funeral practices is probably going to be disappointed and confused. However it may also be that 'new creations' demand new genres and styles, new wine skins for new wine, so that only by the new 'history-like' genre of 'gospel' can the act of revelation be known.[96]

Torrance would however dissent from Ricoeur's elevation of the poetical over the referential functions of text – the poetic function not being understood by Ricoeur as one standing alongside others, such as prophecy, prescription, etc. but as decisively present across all of them:

> My deepest conviction is that poetic language alone restores to us that participation in or belonging to an order of things which precedes our capacity to oppose ourselves to things taken as objects opposed to a subject. …
>
> I am first defining the poetic function in a negative manner, following Roman Jakobson, as the inverse of the referential function understood in a narrow descriptive sense, then in a positive way as what in my volume on metaphor I call the metaphorical sense.[97]

Conversely Torrance[98] is concerned with the scientific function of language, rather than the poetic or rhetorical, the latter of which, like Avis (1976), he finds predominantly restricted to the hymnic Biblical texts. Moore also protests at the arbitrary, homogenising and systematic elevation of the poetic function in Ricoeur.[99] Torrance would also no doubt criticise Ricoeur for his yielding of too much of theology's ground to philosophy, particularly in Ricoeur's insistence that he is to be converted as philosopher within philosophy.[100]

---

[95] For other examples see Jeanrond, 1994, p. 87.
[96] Jeanrond, 1994, p. 90.
[97] Ricoeur, 1980, p. 101, also p. 100.
[98] If one is going to accept that there is such a disjunction between the poetic and the referential.
[99] Moore, 2003, p. 224.
[100] Ricoeur, 1981, p. 157.

Torrance's overarching concern for the scientific reference of the Biblical texts makes his integration into traditional hermeneutical categories rather tangential. His basic critique is that most hermeneutics tend to pose questions prior to or abstracted from an actual text, whereas, analogous to the case of scientific method – indeed given the hermeneutical nature of all sciences, *hermeneutical questions are now questions of scientific method* – these must be answered immanently from within our engagement with the text.

Different books of the Bible are therefore liable to yield different answers on questions of hermeneutical method, a sign of which is the fact that each school of hermeneutics possesses its particular favourites within the canon, namely those which best illustrate its view point. Is the author's intention paramount? The answer to that question is likely to be different depending on whether one is reading an autobiography or a piece of romantic poetry (for instance). Does the text have fundamentally one meaning or many? Is there a sub-text which subverts the text? Is the text coherent, and does it aim to be so? Does the text seek to refer to events in the natural, historical or social orders? Does the text call for a response from the reader, and what sort of response is it trying to illicit? It may be exhorting him to give his life to Jesus, summoning him to appear before the magistrate, or instructing him on the assembly of flat-pack furniture. How does this text refer both internally and to other texts? To what extent are the language and conceptualities embedded within the text affected by 'life-worlds' prior, contemporary and subsequent to the text? Crucially is this testimony, sincere, competent and reliable? It may be a hoax or a joke. And the first question must always therefore be: what sort of text have we here? All these questions are only ones which can be put to individual texts as we read them, they cannot be answered in advance.

> It is necessary therefore to enter the hermeneutical circle. Only in the understanding of the text do I in fact know the object. Faith in what the text is concerned with must be deciphered in the text which speaks of it, and in the confession of faith of the primitive church which is expressed in the text. This is why there is a circle: to understand the text, it is necessary to believe in what the text announces to me; but what the text announces to me is given nowhere but in the text. This is why it is necessary to understand the text in order to believe.[101]

So the hermeneutical circle of understanding entails that in order to understand any text, certain insights have necessarily to be formed

---

[101] Ricoeur, 1981, p. 59.

about the text, which cannot be gained apart from reading of the text – the axiom of genre must be initially held fluidly as an unfamiliar text may spring surprises on us.

Thus Torrance is left with two outcomes. Firstly, given that the answers to questions that would most aid us in our understanding and interpretation of a text are found within the text itself, this in turn implies that the hermeneutical circle is deeply embedded within all our understanding, no less in those texts which refer to themselves as 'scientific'. Secondly, Torrance like Ricoeur, understands that once we have employed a hermeneutic of suspicion (explanation), we need to enter into a second naiveté, where that suspicion will be balanced with a hermeneutic of restoration (descriptive interpretation):[102]

> One has to resist any attempt to reduce religious language to ideology, to resentment, or to obsessive neurosis, as Marx, Nietzsche and Freud asked us to do. …
>
> Hermeneutics seems to me to be animated by this double motivation: *willingness to suspect, willingness to listen: vow of rigor, vow of obedience.* In our time we have not finished doing away with *idols* and we have barely begun to listen to *symbols.*[103]

This is not only a matter of a balanced hermeneutical method, but one that prevents the hermeneutic of suspicion itself becoming a truth-obscuring idol, in which as Thiselton puts it, "the use of a hermeneutic of suspicion as a method" is transposed into a 'world view' determined by suspicion.[104] As we have previously appended Feyerabend's name to the traditional trio of hermeneuts of suspicion, he therefore equally also falls under Ricoeur's, Thiselton's, and Torrance's censure.

For it is this closing of the circle which Feyerabend fails to achieve. His hermeneutic of suspicion with regard to natural science, is well developed, but he does not close the circle with the asking of the question: 'What does all this amount to?' There is no balancing 'vow of obedience'. If conversely Torrance overplays the vow of obedience, this must be correlate with the fact that the Scriptures act as witness to the events of revelation which of their nature demand obedience, and that our suspicion of them must only be the legitimate suspicion of all human record. They may of course narrate an event which is in fact not

[102] Frei, 1992, p. 12.
[103] Ricoeur, 1981, p. 43; *Freud and Philosophy*, p. 27.
[104] Thiselton, 1992, p. 126.

revelatory at all, but that can only be known by examination of that event itself, and is not signalled by any qualities, be they weaknesses or indeed strengths, of the witness.

This question of ideological reading has dogged hermeneutics. Notions of a neutral[105] 'ideal' reader have been abandoned – for even as a limit point natural interpretations are liable to mean that we construct such a reader in our own image, and hermeneuts have generally been more subtle than Feyerabend in not equating this plurality of readers with too easy a relativism –just as quantum and relativity theory have denied us the impossibility of a physically neutral observer. Theological readings must also hold these tensions, with texts whose referent possesses an utter subjectivity which cannot be drawn so easily and equally into a dialectical relationship with us as readers. Thus here even our scientific attempts at impartiality as subjects are relativised by the derivity of our position as subjects. Whether God's sovereignty therefore *needs* to be guarded as jealously *by us* as Barth and Torrance seem to deem necessary is an open question, but we are forced to posit a tension between theological and philosophical hermeneutics, both of which stand quite apart from Feyerabend's position. The demand to be converted as a philosopher, if it is to mean anything, must offer exemption or at least priority to some significant portion of that *philosophy* within the dialogue of conversion. God's sovereignty, even down to the notion of non-contradiction, according to Barth, can grant no such exemption.

Thus the history of the ideal reader is analogous to the ideal observer in natural science, in that developments internal to hermeneutics and natural science have uncovered the dangers as well as the utility of an ideal which can only be regarded as an ambiguous fiction, known only as a point on the horizon, but always incarnated as an historical and personal reader / observer striving for scientific impartiality.

[105] Thiselton convincingly distinguishes between 'neutral' and 'impartial' readings, analogous to the role of a home team referee in a sporting contest, who cannot claim to be neutral, but can still strive to be impartial. Thus a declaration of *non*-neutrality, of interest, is not in conflict with impartiality, but is a necessary preliminary to it (Thiselton, 1992, p. 444. Richard Coggins in his final essay in the collection *Open Text* (Watson (ed.), 1993), absolutely misses this distinction, and uses the terms as though they were interchangeable.) Torrance's reading of the text, as an 'objective scientific' reading, is impartial with regard to faithfulness to object, but it cannot be a 'neutral' reading – precisely in proportion to the fact that the text is itself not 'neutral' but a confession of faith (Torrance, 1969, p. 303).

* * *

Torrance and Feyerabend both therefore understand the hermeneutical circle of understanding to be universally present in all acts of human knowing, including those which claim for themselves the predicate of scientific. For instance our natural scientific attempt to depict reality is carried out within the (non vicious) circularity of theory and observation.[106] Gadamer's and Heidegger's description of the role of preunderstanding in hermeneutics, has been shown by Feyerabend not to be absent from the natural science, when similarly, established theoretic frameworks are likely to endeavour to manipulate the 'text' of observational data in a conservative direction. The situation is more similar than we might imagine across the disciplines, and Betti's criticism of Gadamer's loss of objectivity cannot therefore be given much weight.[107]

However there are substantial differences as to how that circle functions, depending upon which side of the socio-critical and socio-pragmatic hermeneutical dividing line it falls upon. Torrance for theo-anthropological reasons, at least as far as the Biblical material is concerned, contends that we stand more under its judgement than it under ours. This makes his a socio-critical reading, in that the dynamic of proclamation is more of a challenge to us than it is a challenge of suspicion towards that proclamation, though critical and scientific procedure demands that the latter is properly present within our bearing towards the received Biblical text.

The balance of power, according to Feyerabend, should lie with the other pole, the pole of human knowing. Ultimately what knowledge we have depends upon us as mature human subjects:

> One might call the omnipresence of this choice the 'existential dimension' of research ... The questions 'what shall we do?', 'how shall we proceed?' 'what rules shall we adopt?' 'What standards are there to guide us?' however, are answered by saying: 'you are grown up, children, and so you have to find your own way.'[108]

Questions of scientific method or hermeneutical procedure are not questions about epistemology at all, but moral or even aesthetic choices.[109]

---

[106] Smedes, 2004, p. 94, who also quotes Drees, *Religion, science and naturalism*, 1996, p. 132f.

[107] Pannenberg, 1976, p. 165f.

[108] Feyerabend, RRSM, p. 19.

[109] Preston, 1997, p. 20–21.

Both Feyerabend's and Torrance's hermeneutics demand that in reading the text an existential decision is demanded of the reader.[110] For both of them the decision is existential, in that the text demands more than offering our assent to a statement like '2 × 2 = 4' – it is for both of them a real and substantial decision.[111]

For Feyerabend the reader stands in ethical judgement *over* the text, for Torrance we stand *under* the ethical judgement *of* the text (the existential decision is whether we submit to its judgement – the Word of God as addressed to us). In Feyerabend the text seeks from us approval and employment, in Torrance the text seeks our submission. Feyerabend's reading of the text is therefore socio-pragmatic (for instance in his previously cited approval of liberation theology) rather than socio-critical.

Here Feyerabend simply has not balanced his hermeneutic of suspicion with a vow of obedience, as Ricoeur would point out, and it has therefore turned, as Thiselton says from a method into a world-view. Whilst both Torrance and Feyerabend refuse to demarcate science from hermeneutics, but point with Gadamer to the hermeneutical components within all sciences, Feyerabend's unwarranted further identification of the sciences as producing 'works of art', is a sign that his radicalised version of this insight has not succeeded – the referential functions of scientific knowledge, however ambiguous they might be, cannot simply be reduced to the level of importance they have in art.

Whilst Comte and Mill,[112] the later being Feyerabend's hero, might try and align human and natural scientific methods, Dilthey is correct in (particular within his own historical context) to insist that methods are formulated corresponding to the nature of the object, as Torrance would wish. However in then dividing those methods between the mathematical and the hermeneutical sciences, the danger was that, thereby the human sciences were liable to be judged of inferior merit to the 'exact' sciences, particularly in the scientistic climate of the time.

[110] "We are confronted with a real *decision*, that is a real choice, with a situation which has to be resolved on the basis of our demands and preferences, and which cannot be resolved by truth." Feyerabend, 1961, p. 56: "The problem of responsible choice enters even the most abstract philosophical matters and … ethics is, therefore, the basis of everything else." Feyerabend, 1965, *Problems of empiricism*, p. 219, quoted by Preston, 1997, p. 21.

[111] "The act of obedience we may describe as *decision*. No decision is required when we state that 2×2=4. We are concerned there with what is timeless and necessary … no choice, no decision is involved … In theology however, we cannot discern the truth except through decision." Torrance, 1969, p. 214.

[112] Schmidt, 2006, p. 33.

Gadamer, in unveiling the interpretative pole of *all* sciences, (though this does not necessarily contradict Dilthey, who like Ricoeur,[113] was keen to underline the dialectical continuities between *verstehen* and *erklären*) underlines the recent developments in modern physics which highlight the need for careful consideration of differing interpretative and philosophical commitments. These have proved as important as empirical and mathematical work – indeed there has been a cross-fertilisation between them. This then allows the human sciences a corresponding recovery in epistemic pride which had been severely dented by scientistic and positivistic influences.

Realism

If Torrance's hermeneutics is driven by the majesty of its referent, then the manner in which language obtains, shapes, or perhaps even creates such reference, is of considerable importance. Torrance's distrust of nominalism[114] betrays what Hardy terms "the most highly developed version of realism" available in modern theology.[115] McGrath[116] is quite clear at this point that Torrance is using the term realism as analogous within both natural and theological sciences, so that Einstein and Maxwell are offered as examples of realists whom theologians might imitate, and not as later used in a specifically theological context by Cupitt and his followers. McGrath designates Torrance's realism as *scientific realism,*[117] following the approach of Kukla, Psillos and Boyd.

Van Fraassen (for example), who argues defines scientific realism as:

> Science aims to give us, in its theories, a literally true story of what the world is like; and acceptance of a scientific theory involves the belief that it is true. This is the correct statement of scientific realism.

As opposed to his own position, a non-realist constructive empiricist view of scientific theory:

---

[113] Jeanrond, 1994, p. 73.

[114] Torrance, 1982, p. 49, though Torrance's identification of nominalism as an inherent part of the scholastic outlook is incorrect.

[115] See McGrath, 1998, p. 211 – Patterson also identifies Torrance as paradigmatic 'contemporary theological realist' operating with a 'linguistic-window-on-reality model' – 1999, p. 1.

[116] McGrath himself will develop this in greater detail in 2002, Vol. 2, p. 121–end.

[117] McGrath, 2002, p. 130.

> Science aims to give us theories which are empirically adequate: and acceptance of a theory involves as belief only that it is empirically adequate. This is the statement of the anti-realist position I advocate; I shall call it constructive empiricism.[118]

Torrance opts for the former view in both theological and natural science. Theology's task is to speak truthfully about the reality of God, and not simply to order the data in an 'empirically' adequate manner, even if knowledge of that reality is mediated through human language and experience. Thus the incarnational centre of Torrance's theology, as the point wherein God is revealed in human flesh, is combined with Torrance's rejection of any attempt here to interpose an economic (Torrance prefers *evangelical*) Trinity in contradistinction to an essential (Torrance again prefers *theological*[119]) one, to make a very direct link between the actual words of Jesus Christ, and his reality as God: "What God is toward us in his historical self-manifestation to us in the Gospel as Father, Son and Holy Spirit, he revealed to be inherently and eternally in himself."[120]

As we have seen, Feyerabend's ultimate position is deeply ambiguous in relation to the question of scientific realism. There is certainly no worked through programme of any sort of non-realism. His journey away from realism must be understood principally as a reaction against his own earlier super-realism,[121, 122] and the belief that, given the confusion, it would be better if ethical considerations gained priority – science if humbled must avoid pretension. As we have also seen, his

---

[118] Van Fraassen, 1980, p. 8, 12.

[119] Torrance, 1996, p. 7; we note incidentally that this terminology stemmed from an ancient custom dating to the third century by which what we now know as the discipline of theology was divided into two: theology being a science of God in its proper sense, and economy being the science of God's dealings with humanity (Pannenberg, 1976, p. 7).

[120] Torrance, 1996, p. 1.

[121] Feyerabend, RRSM, p. 158.

[122] Super-realism is the belief that the solution to the riddle of Eddington's table (see A. E. Eddington, *The nature of the physical world*, CUP, 1928, p. xif) lays in our abandonment of common sense notions of 'table' and replacing them with a more thoroughly 'scientific' account. Feyerabend's analogy with other words which lose their reference to reality (angels, demons) does not work, because such terms do not refer to realities otherwise referred to in the material world, but refer to another super-natural spiritual world ('material' and 'spiritual' are terms employed by Feyerabend). From 1957 Feyerabend develops Hollitscher's argument that realism is a founding assumption of scientific research, not a result, whereas positivism would necessarily stultify such research (Preston, 1997, p. 31). His study of quantum mechanics leads him to the conclusion that ultimately the inadequacies of quantum theory must eventually give way to a more complete theory.

incommensurability thesis is robbed of its potency if the demands of realism do not keep the ontologies of incommensurable theories rubbing up against one another.[123] As he develops his incommensurability thesis and slides ever closer to epistemic relativism, so his realism gives way. Scientific paradigms are no longer to be judged on their merits as descriptions of reality but as part of traditions to be assessed on their humanitarian credentials and pragmatic benefits.

Scientific realism[124] is a jealous god, and demands absolute priority of representation, be it theological or physical. This search for a true correspondence between statements and the realities to which they refer, takes on in Torrance, its own sense of moral imperative – scientific discipline entails the ethical imperative to tell the truth about reality, whether or not that truth has uncomfortable or inconvenient consequences for human beings. Feyerabend never positively argues *against* such realist views of science on epistemic grounds – it is simply that *given* the logical disjunction between incommensurable traditions the most vital criteria *left are* ethical. The demand that theory should conform to some unbending external reality is for Feyerabend inevitably at the mercy of a pathological scientistic imperialism, whose account of the 'facts' of that reality is already riddled with their theories - preparatory groundwork for a therefore one-sided imposition of a 'rational' and 'objective' debate.

Feyerabend's liberal optimism insists that it is better to *choose* a reality proven to enable human flourishing than try and refine our theories and conceptualities to conform with the facts as best we know them, theory laden as they must be. Torrance's Calvinism entails a certain wariness of human choice as a reliable guide, preferring to insist that we try and approach the true nature of our object by increasingly coordinating our statements to the reality of that object, even if such coordination can ever only proceed by approximation and iterative fluidity, not to mention that the attempt will not be immune from the effects of ordinary human frailty.

At the time of *Against Method*, Feyerabend seems be passing through the territory of critical realism on his journey from super-realist to non-realist. He quotes Ziman's[125] map metaphor:

---

[123] Burian, 1984.

[124] i.e. as per van Fraassen, its theories are *true*.

[125] Feyerabend AM, p. 245, reference is to J. Ziman, *Teaching and learning about science and society*, 1980, p. 19 – though Feyerabend has an odd idea that we always correct maps as we go, and we'll soon 'get into trouble' if we use the map 'no matter what' – AM, p. 223 – this is not mine, nor I suspect many people's experiences as walkers.

> There is no simple 'scientific' map of reality – or if there were, it would be much too complicated and unwieldy to be grasped or used by anyone. But there are many different maps of reality, from a variety of scientific viewpoints.

Obviously reality is still a category here, even if we can only communicate its nature through scientific maps. This critical realism is the sort of realism which McGrath adopts, and which he detects in Torrance,[126] also citing the work of Achtemeier, MacIntyre, N. T. Wright, J. D. Morison and Polanyi, though in later work McGrath discovers the work of Roy Bhaskar[127] as singularly exemplary of this position.[128]

McGrath's description of Torrance's critical realism is as follows:

> A central assumption of both the natural sciences and of scientific theology is that that there is indeed a 'reality' independent of the human mind, which the mind is capable of grasping, while fully conceding that:
>
> 1. Our grasp of reality is always less than total,
> 2. our descriptions of reality must be held to be revisable, in the light of closer approximations to the essential properties of things,
> 3. models or other constructs may be required when representing or analysing this reality, or its aspects, without in any way calling that reality into question.
>
> This belief is not regarded as an *a priori* dogma, but as an outcome of an engagement with the world.[129]

This is a description of Torrance's critical realism which I would wish to endorse, both as description and as philosophy.

* * *

In a more contemporary discussion of realism, Andrew Moore[130] distinguishes between three elements of realism – ontological (which

[126] McGrath, 1999, p. 217.

[127] McGrath, 2004, p. 139ff.

[128] Bhaskar in Archer et al. (eds.), 1998, p. 596; see also Bhaskar, 1991, p. 77.

[129] McGrath, 2002 Vol. 2, p. 122.

[130] Moore, 2003, p. 1; Michael Dummett (1978, p. xxxi) too expresses concerns over the coherence of a unitary concept 'realism' ("it was obscure whether the term had a unitary meaning"), and contrasts the "two quite different pairs realism / nominalism and realism / phenomenalism." However despite the contrast in his original paper (p. 145f) Dummett explains that though whilst at school he was taught that the medieval dispute between realists and nominalists had nothing to do with later philosophical disputes between realism and idealists, it was the subject matter which was most divergent, and that in fact there was a significant "resemblance in the form

opposes idealism), epistemological (in opposition to empiricism) and semantic (in opposition to linguistic idealism).[131] These three are interconnected, and though not always so subsumed, the latter tend to embrace the former.[132] It is also the case that they are not always easily or systematically distinguished. As Westphal elegantly points out,[133] the problem is that the 'paradigmatic anti-realist' is Kant, and Kant is a 'metaphysical realist' and 'theist'. Furthermore Kant therefore believes that the 'God's eye view' is real, but inaccessible, whereas non-theistic philosophers such as Rorty and Feyerabend radicalise this view, so that once God is eliminated, so is his 'view' and hence any metaphysical absolute. The debate concerning realism becomes therefore a matter of epistemic degrees,[134] between 'creative' anti-realists on one hand (labelled thus by Plantinga, e.g. Feyerabend) who emphasize the relative importance of human decisions within knowledge since the Kantian Copernican revolution, and critical realists who emphasize the metaphysical constraints that reality imposes upon knowledge (i.e. Torrance). Unless we define our terms in some detail it is likely that we shall fall foul of Nielsen's charge that "both metaphysical realism and anti-realism are metaphysical constructs which take in each other's dirty linen."[135]

Although Moore's concern is with ontological[136] realism in a particularly Christocentric form (like Torrance), his focus is mainly epistemic (and also, though less so, semantic), as the condition and assumption required by ontological realism,[137] therefore his argument

of the disputes." His useful definition (p. 146) runs thus: "Realism ... is the belief that statements of the disputed class possess an objective truth value, independently of our means of knowing it: they are true or false in virtue of a reality existing independently of us. The anti-realist opposes to this the view that statements of the disputed class are to be understood only by reference to the sort of things which we count as evidence of a statement of that class."

[131] See also Hacking, 2000, p. 194.

[132] The relation between the latter two depends on to a large extent on whether, like Torrance, we concur with Polyani's maxim "we know more than we can tell." (e.g. Polanyi, 1967, p. 4).

[133] in Moore & Scott (eds.), 2007, p. 131f.

[134] Westphal in Moore & Scott (eds.), 2007, p. 132.

[135] Nielsen in Nielsen & Phillips, 2005, p. 195.

[136] Moore, 2003, p. 7.

[137] "Religious realists typically maintain that religious claims represent a mind-independent religious reality to which we have epistemic access (at least in part), and that religious truth should be robustly constructed as a relationship between religious sentences and the reality that they describe. Religious realists also usually maintain

is a 'transcendental' one[138] (again like Torrance). This is, particularly in his top-down Christocentric emphasis, close to Torrance's approach.

However, Moore is consistently critical of any analogical equivalence between scientific realism (which is generally ontological and epistemic[139]) and theological realism. Partly this is because like Torrance, he prefers to talk of Christian realism to theological realism,[140] but that is internal to the theological pole of the analogy, it does not pertain to the analogy itself. I therefore believe that Moore makes the following errors in his judgements about the possibility of the analogy.[141]

Firstly, Moore highlights the use of the analogy between unobservables in natural science, and the unobservability of God.[142] If (as he accuses) Soskice[143] and others depend on this analogy – that the analogy 'has a core of observability'[144] – then that is their mistake. My analogy (and Torrance's) is between theoretic terms in natural and theological science. When Moore notes the Scripture's ambiguity over the observability of God (we may see his glory, but not his essence), this may well not be analogous with unobservables (such as electrons), but it is with theoretical terms such as force – we may not be able to see

that at least some religious claims are actually true. Religious anti-realists variously reject different components of the realists' theory: religious claims are primarily expressive rather than representational; religious truths are inaccessible to us; religious truth is a matter of the satisfaction of internal standards of religious language (or 'language games'); religious claims are systematically false." Moore & Scott in Moore & Scott (eds.), 2007, p. 1.

In Torrance's Scottish inheritance this is an argument put forward by Edward Caird (see Caird, *The evolution of religion*, in Fergusson (ed.), 2007, p. 138).

138 Moore, 2003, p. 14.

139 Bird in Moore & Scott (eds.), 2007, p. 61f.

140 Moore, 2003, p. 40f.

141 Also he seems to be rather hard on 'theological realism' and the 'god of the philosophers' (p. 30) – harder even than Barth and Torrance. Certainly the 'god of the philosophers' is an anaemic being, compared to the triune Biblical God, but he is a shadow not a rival. Certainly the picture needs filling out, but not necessarily rubbing out. So Barth and Torrance's (Moore himself says this, p. 32) lack of enthusiasm for the ontological proof of God, is lack of enthusiasm for its role as a 'proof' in the Cartesian sense, not for its Anselmic role in the *Proslogion* 'addressed to God as prayer'. Certainly theological realism, if built up from 'religious experience', is not as powerful as 'Christian realism' built down from revelation, but it is unconvincing rather than idolatrous.

142 Moore, 2003, p. 11.

143 Moore, 2003, 2003, particularly p. 40f.

144 Moore, 2003, p. 43.

a field of force, in essence, but we may verify – see, feel, hear, – its 'glory' (effects).[145]

Secondly, he notes other dissimilarities within the analogy – well that is the point of analogies – to say as he does, that the analogy does not hold, is to maintain that some sort of tipping point has been reached. Furthermore those dissimilarities (novel predictions, repeatability of experiments[146] in particular) are found principally within the inductive physical sciences, whereas we have already made the case for theology being an abductive science, alongside others: physical, historical and social. Torrance's second particularity of theological science – that it is of a person – implies that it will quite properly be ambiguous with regard to predictability and repeatability. However other possibilities open up in the field of onto-relations – such as a *deepening* of personal relationships – this is denied natural scientific confirmation where verification is *designed to be* a pretty binary kind of exercise.

Following his engagement with natural science, Torrance is (unlike Moore) therefore committed to an analogy between Christian realism (not simply theological realism) and scientific realism (McGrath's second volume of *Scientific Theology* is predicated upon this analogy).[147] The question is therefore whether the 'tipping point' has been reached – or whether, as I would maintain, Torrance's explication of similarities and differences has sufficiently cleared up confusions of which the analogy might be the cause.

Of course the substantive literature of Torrance and Feyerabend comes from an era (roughly 1969–85) when such distinctions within realism were not so clearly made. Certainly the radical theological ontological anti-realism of Cupitt and his followers (the Sea of Faith Network) comes after this period (beginning with the publication of Cupitt's *Taking leave of God* (1980) and the Anthony Freeman case (early 1990s)), and though the connections are clearly present, deals with rather different issues. That is, of course, not to say that Torrance is in any sense ambiguous about ontological realism, but that his form of epistemic realism, and his desire to maintain the scientific prevenience of God, prevents him from *contending directly* for such realism.

---

145 Moore, 2003, p. 67–8.

146 Moore, 2003, p. 50.

147 See also Smedes, 2004, p. 92, who identifies Polkinghorne as also using the same analogy.

Correspondingly, we note that Cupitt is similarly uninterested in giving warranted proofs to ontological questions about God's existence. Cupitt's detractors therefore also often miss their target. Certainly Cupitt points out[148] and catalogues the problems with the traditional theistic proofs of God's existence, but Torrance equally distrusts such arguments and thus rather draws Cupitt's sting at this point.[149] Equally Cupitt is not particularly interested in advancing proofs of God's non-existence.

Rather what Cupitt offers, is a refreshingly candid account of what Christian faith might look like for those for whom, for whatever reason, claims for God's existence as transcendent, cosmic and personal *have become* untenable. This is therefore actually quite similar to Torrance, who offers systematic theology as an ordered reflection on the components of Christian revelation, for those who, again for whatever reason, are persuaded of a more orthodox understanding of that revelation and therefore of the doctrine of God's being and existence. Like a mirror image of Cupitt, Torrance's theology is very much a conditional and posterior one. Cupitt shrinks from responding directly to his conservative critics, suffice to show that the arguments of classical theism are weak. Instead he prefers to say that *if* classical theism is no longer tenable, then here is a version of Christian Buddhism[150] which may fit the bill. Torrance in the same way, says *if* [a real] God has (objectively) revealed himself in Jesus Christ, then here is what systematic theology might look like. The defence of Torrance's conditional is largely fideistic (indeed personal) and coherentist (though none the less realist for that). Torrance does not shrink from considering whether such fideism might just be, as Pannenberg says, illusion and caprice; but argues that the nature of the object itself implies that its justification is interior and posterior to the ongoing engagement with that object in Christian worship, life and theology – there is no way around by which we might satisfy Enlightenment prejudice against prejudice.

However Cupitt's failure to distinguish properly between epistemic and ontological issues[151] leads him to base his theological non-realism

---

[148] Cupitt. 1980, p. 20ff.

[149] Cupitt, 1980, p. 7, cites Barth, along with Tertullian, Luther, Pascal and Kierkegaard on precisely this point.

[150] Cupitt, 1980, p. xii.

[151] McGrath, 2002, p. 253.

squarely on the basis of a rather crude constructivist scientific non-realism:

> Does this amount to saying that God is simply a humanly constructed ideal, such that when there are human beings no longer, there will be God no longer? The question is improper, because it is framed from the obsolete realist point of view. The suggestion that the idea of God is man-made would only seem startling if we could point by contrast to something that has not been made by men. But since our thought shapes all its objects, we cannot. In an innocuous sense, all our normative ideas have been posited by ourselves, including the truths of logic and mathematics as well as our ideals and values. How else could we have acquired them? Thus God is man-made only in the sense that everything is.[152]

However this is incoherent, for when it suits his purposes – for instance when considering Darwin's theory of evolution and its challenge to Christian theology[153] – he is capable of taking a straightforwardly realist view of scientific theories. Whilst it is perfectly reasonable to be a natural scientific realist and theological non-realist, Cupitt cannot then use overarching social constructivist arguments – either theology and natural science are in the same boat when it comes to realism, or they are not – he cannot change tack mid-argument to suit his needs.

Feyerabend's 'non-realism' is, as noted earlier, more a consequence of his dropping of his earlier 'super-realism' than a worked out position. Just as he is always skipping between different sorts of relativism, so he skips around different forms of non-realism.[154] Sometimes he speaks of reality[155] or facts, as being chosen[156] or sculpted[157] by human knowledge – that, however unguardedly, sounds like ontological non-realism. Certainly his appropriation of theories of language by Whorf and Wittgenstein implies that there is a strong element of semantic non-realism – though we by no means endorse such an interpretation of Wittgenstein – and though he shares similar concerns, there is no citation of deconstructionist or post-structuralist schools of thought. However, given that ultimately his interest in linguistics is secondary and subservient to his primary one as a philosopher of science, it is

---

152 Cupitt, 1985, p. 270.
153 Cupitt, 1985, p. 64f.
154 Worrall, 1991.
155 Feyerabend, FTR, p. 73.
156 See Feyerabend, RRSM, P. 19.
157 See Feyerabend, AM, p. 269.

scientific realism, and therefore the epistemic component of realism, with which he is ultimately concerned. When he talks of 'traditions', he is referring to culturally determined and pragmatically orientated ways of seeing a visceral reality, the ontological reality of which is not, except in rather throw-away passages, open to question. However the danger of such throw away passages is great: the Sokal hoax,[158] depended, prior to the more general nonsense in the body of the argument, on an initial coalescence of epistemic and ontological non-realism, which others took up uncritically.[159]

In his employment of 'traditions', Feyerabend is reminding us that natural science, and science in general, is itself traditioned. What he wishes to rebut, is scientists trying to portray their tradition as rising like a gleaming jewel of objectivity from the ignorant mire of other efforts at knowledge within the pantheon of human traditions. However it might equally well be argued that it is precisely by the coming to consciousness of this insight – that there are traditioned elements, past and present within all human truth-seeking enterprises – that we become able to denote as 'scientific' those disciplines which have become so aware, and therefore are enabled to develop strategies to allow for or transcend them – thus putatively offering an escape from the tradition-determined circle by means of the tradition-mediated spiral. Thus this self-awareness is itself a first step towards an objectivity which transcends tradition, because in acknowledging its traditioned status it takes a leap which most traditions do not. Ironically Feyerabend's critique of Western notions of science, far from deconstructing them, has provided an important tool aspect of self-awareness, necessary within the search for objectivity.

Thus with regard to Feyerabend's challenge to Torrance, it is therefore generally questions of scientific and epistemic realism with which we are concerned.[160] Of course that does not rule out ontological or

---

[158] See A. Sokal & J. Bricmont, *Fashionable nonsense: Intellectuals' abuse of science*, New York, Picador, 1998; see e.g. McGrath, 2002, p. 188.

[159] Moore, 2003, p. 178.

[160] Despite this disclaimer one must give some passing account of non-realism as we encounter it in Cupitt and the Sea of Faith Network (see Moore, 2003, p. 112f, p. 194f). Torrance would no doubt say that given one cannot prove *a priori* the existence of God, there can be no direct rebuttal of such a position. Non-realists typically point out that in terms of prayer, worship, practices, ethics, scriptures, even the grammar of doctrinal discussion, they have no arguments with the realists (Moore, 2003, p. 87f). They are quite content to be in the same church with them, as long as other members of that church respect the fact that when they as non-realsits talk of God – they are

semantic issues – indeed perhaps we might construct a coherent continuum of realisms and non-realisms, from radical idealism to naïve realism.[161] This would have the advantage of identifying those frequent times, when realists and non-realists argue past each other, in the rush for the middle ground – quite often realists rebut idealist positions, whereas empiricists have no particular care for either. The precise construction of such a continuum is beyond the scope of this project, so we have restricted ourselves to the sort of realism with which Torrance and Feyerabend are concerned. It seems to me, furthermore, that Ian Hacking's juxtaposition of 'nominalist' versus 'inherent structuredness'[162] brings out with sufficient clarity, but also room for manoeuvre, the issues involved in scientific realism. Obviously nominalist is here employed in a contemporary and scientific manner, rather than in a scholastic theological sense – though such resonances may be

---

not referring to an independent objective Deity, but to something which comes from within human beings, and represents their highest values and hopes. If intolerant Bishops (as in Anthony Freeman's case) choose to remove popular parish priests from their posts, against the will of many of their parishioners, then that is simply a manifestation of their narrow mindedness. Torrance, would no doubt see them as, basically wasting their time – they have mistaken the essential reference of faith (the gulf of sense is too big to be crossed by any reasonable charity of reference), and with it the whole integrity of the theological enterprise – whether they have succeeded in creating a self-sustaining and coherent fairy-tale is of course another matter. However a more serious question for Cupitt, would be how such a faith would guard against projecting, not simply the finest and the best of humanity, but the most sinful and the degrading, albeit unconsciously, upon their 'God'.

Therefore my own personal answer to such non-realists, if I might be allowed it, is similarly not an intellectual rebuttal, but a few pragmatic observations. Firstly I would draw attention to the fact that as Moore says, only 'philosophers and eccentrics' worry themselves unduly over the basis of realism (Moore, 2003, p. 194). Secondly and relatedly, such theological non-realism is always, to be blunt, so unremittingly high-brow. Talk of God as representing the 'highest values of human beings', always seems to amount to talking about Kant, Plato, Bach, Goethe, Proust, Picasso in preference to Eastenders and the X-Factor. As a result of which I doubt whether the 'common religious practises', demanded by the non-realist God, will be all that common any more. The defence left to the network is to say that it is precisely that – a 'network' – not a church – and that such things as mission belong to the whole church.

Moore gives perhaps the better and less context-specific example of martyrs (Moore, 2003, p. 195f). The Christian faith sets great store by the witness of the martyrs, who one imagines were a fairly realist bunch. It's by no means impossible to imagine, but many would need convincing that a non-realist theology has enough depth to strengthen and uphold many of its followers at the place of execution. The non-realist God is not 'the Almighty' of religious discourse, being too dependent on the human intellects of which he is a function.

[161] As e.g. Smedes does, 2004, p. 94.

[162] Hacking, 2000, p. 83.

helpful being concerned with issues "somewhat though not wholly removed."[163]

Furthermore it should be noted that the form of realism is also determined by the object of study.[164] This is crucial to critical realism and the new objectivity. Thus realism about material objects is opposed to phenomenalism; realism about theoretical entities is opposed to positivism; realism about mathematical statements (Platonism) is opposed to constructivism; realism about mental states is opposed to behaviourism. Thus theological realism, being about God, will have its own particularities. A mark of *critical* realism is that it has a sensitive regard to the distinctive nature of each supposed object. Furthermore, as long as we avoid undue extrapolations from one object to another, it is not necessarily incoherent to be, for instance, a phenomenalist and a mathematical Platonist. To be a realist about one set of objects does not necessarily commit us to realism with regard to another, unless such a realism is justified in terms (for instance around a general theory about the relation between signifiers and signified) which transcend that particular category of objects. Thus whilst Torrance may find the analogous realism in natural science and theology illuminating, convenient and arresting, there can be no question of entailment – rather they are simply both basking in the mutually conducive climate. However Luoma argues that Torrance has fallen precisely into the trap of simply and naively lumping together a series of discontinuous arguments about realism.[165]

According to Torrance, in the relations between reality at an ontic level, our perception of it, our mathematical quantification of it, our linguistic description of it,[166] between signifier and signified, as creatures we are bound to encounter an uncrossable linguistic and logical ditch: "It is impossible to say in language how language is related to being. Belief that this can or ought to be done was one of the root errors of medieval epistemology and hermeneutics."[167] We cannot express in mathematical terms alone the way in which mathematics is related to physical reality,[168] and therefore cannot legitimise that

[163] Moore, 2003, p. 2 (footnote).
[164] Dummett, 1978, p. 147, 367.
[165] Luoma, 2002, p. 63.
[166] Wittgenstein, 2001, prop 4.01, (in fact it is 4.12 which is referred to by Torrance in this regard, 1969, p. 24, 183).
[167] Torrance, 1965, p. 92.
[168] Wittgenstein, 2001, prop 1.211.

relation within a closed system – that is the 'unreasonable' component of mathematics' effectiveness, as Wigner terms it.[169] Here the complexities of the continuum are manifest.

A third complexity – beyond what sort of realism, and what sort of subject matter we are considering, is what we might call the 'burden of proof' difference. If people assume realism as a working hypothesis, but freely admit to not being able to justify such an assumption, or at least only to justify it on particular grounds, are they realists or non-realists? Torrance is not a scientific realist in the sense that he believes that realism can be justified on non-theological ground. Rather he is a realist in the sense that such a realism is a necessary assumption of all scientific enterprises, an assumption which only when conjoined with another 'unprovable' – the existence of God – mutually reinforce one another so that transcendentally we are justified in making an inference to best explanation,[170] preferable to the nihilism of various sorts of anti-realism, which might possibly be the only conclusion should we take realism on its own. In the typology of realisms there is therefore yet another division beyond level and object, which is that of justification – realists who think that a belief in realism can be shown to be warranted, and realists who think that it must be assumed (Torrance), the latter in turn often share ground with non-realists (Feyerabend), who are simply more sensitive about granting unwarranted assumptions.

There is also the issue that whilst there may well be an external 'reality', our experience of it comes only via our sense organs, and therefore we must be concerned at our sensory interfaces' ability to render that reality accurately and reliably. Can that reality be objectively depicted, having been grapsed and mediated through sensory phenomena, or does much of what is real lie beyond the detection of those senses? We know that the senses are limited (for instance the eyes only respond to electromagnetic waves over a rather narrow frequency band) and liable to play tricks on us. Kantian non-realists underline the gap between phenomena and things-in-themselves; critical realists admit the mediacy of all our experience and try to factor it in.

So whilst for Torrance there can be no *logical and necessary* bridge between the intellectual constructions of human representation and

[169] See for instance McGrath, 1999, p. 225.
[170] Moore, 2007, p. 75.

the sensible reality of the physical world,[171] there is nonetheless a created and implicit correspondence which relies upon the gracious, ordered, extra-logical rationality of the creator – the bridge is theological rather than logical, but formal logic is for Torrance, already living in the reduced circumstances of 'paper logic,'[172] and must take its place alongside other 'logics.' Any attempt at a logical demonstration of that correspondence not only subverts the basis of such representations – linguistic, visual or mathematical – but also attempts to establish the goodness, intelligibility and orderliness of creation upon grounds other than the goodness of God in his act of creation. Like the indeterminacy principle the absence of a logical bridge is itself a 'theo-logical' corollary of Torrance's doctrine of God, rather than being something we might discover in the future. It is a necessary absence. "For true statements to serve the truth of being, they must themselves fall short of it, and be recognized as such, for they do not possess their truth in themselves but in the reality which they serve."[173]

Thus Torrance's appropriation of Wittgensteinian language (see footnote 171) marks a divergence with Feyerabend in his interpretation of Wittgenstein's theory of language. Though Torrance himself "condemns linguistic philosophy as a … superstitious veneration for ordinary language", this too is, as Patterson says, to misrepresent Wittgenstein, being a "narrowness of Torrance's view of linguistic philosophy."[174] We therefore agree with Patterson that Feyerabend (in approval) and Torrance (in disapproval) both get Wittgenstein wrong, and that in fact as per Patterson and MacDonald, the trio Torrance, Barth and Wittgenstein can here be reconciled. For Torrance, our inability to explain how language depicts reality is nonetheless predicated on an understanding that it can and does – language is a 'form of life', but it is still life in all its (God-given) richness not a nihilistic negation of existence. Feyerabend employs the contextual theory of meaning to more radical ends than Wittgenstein or Torrance, presaging developments in post-modern linguistics, in which he questions whether such correspondence is anything more than wishful, and sometimes imperialistically wishful, thinking.

---

[171] "Just as we are unable to picture how a picture pictures a landscape, so we are unable to put into statements how statements bear upon being", Torrance, 1982, p. 73, similar in 1969, p. 223.

[172] Torrance, 1969, p. 219.

[173] Torrance, 1984, p. 320.

[174] 1999, p. 74, note 1.

Thus Feyerabend is left stuck in a Whorfian bind, in that as a result of his employment of the Sapir-Whorf hypothesis, he says that language "shapes and describes reality."[175] That is true. Feyerabend is employing the thesis as a response to logical positivism's attempt to separate observation from theory by constructing a theory-neutral observation language. However after that particular concern has been addressed (and here we can agree with Feyerabend that such a project was a failure), the more vital and general question is therefore which pole has the upper hand, especially when that language is in the hands of people striving to operate 'scientifically'. If the former 'shaping' pole triumphs, we spiral outwardly towards relativism, and if the latter 'describing' one does, we spiral towards semantic realism. Feyerabend accuses Whorf of holding the former view, namely that language is the main and perhaps only 'shaper of events,'[176] a view which he himself regards as 'too narrow'. However having criticised this fortress of the linguistic idealists, Feyerabend ends the paragraph in the most odd way – with a parenthesis about scientific realism, before opening the next paragraph with an abrupt change of tack to incommensurability. He does not inform us of *his* view of the relations between the poles, but assumes he has introduced enough support for the thesis to engender in us some vague precautions about language, which are actually nothing but old news to critical realists, who have long incorporated insights of the weaker version of Whorf's hypothesis into their frameworks, indeed to the point that such considerations differentiate critical realism from its unsatisfactorily naïve alternatives. If linguistic idealism is 'too narrow', then the malady isn't as severe as the symptoms which Feyerabend is painting, and the medicine or anarchy is too drastic.

Perhaps the closest he gets to clearly stating his position is in a later added 'postscript on relativism'[177] where he says that "scientists are sculptors of reality" – however sculptors are primarily shapers, shapers restricted to be sure by the resistance of their material but nonetheless fundamentally shapers rather than describers – or to use Ford's terminology, 'fashioners' rather than 'finders.'[178]

---

[175] Feyerabend, AM, p. 210.

[176] Feyerabend, AM, p. 210.

[177] See Feyerabend, AM, p. 269.

[178] D. Ford, *System, story, performance*, in A. Hauerwas & L. G. Jones, *Why narrative? Readings in narrative theology*, Eerdmans, Grand Rapids, 1989, p. 195–200 also cited in Patterson, 1999, p. 53 – Ford's 'middle-distance realism' defined as holding both in balance.

Moreover within linguistics itself, Sapir-Whorf (particularly in its stronger version) is a contentious thesis, and others such as Chomsky, underline the 'universals of human thought', the so called 'hard wiring' of basic conceptual structures in the human mind.[179] Being a 'shaper' rather than a 'describer' is, incidentally, Torrance's main charge against Kant.[180]

* * *

Both Torrance and Feyerabend would agree that our constitution as beings determines what sort of knowledge we possess, and how we order it. If for instance, we possessed different sense organs – perhaps ones sensitive to electromagnetic frequencies other than those of the visible spectrum – we would no doubt order our knowledge of the world in different ways. For Feyerabend such constructivism is ultimately, with a socio-pragmatic view of science, dominant. For Torrance, though human knowledge is structured by what and who we are, it is still knowledge of something objective and independent, something which in itself has an inherent structure, only by the respecting of which can true knowledge be achieved. This inherent structure can indeed sometimes disrupt our attempts to organise knowledge according to our own principles and methods, and force us to reconsider our conceptual structures, in order that they might be conformed to the nature of reality.

> Belief is not something that is freely chosen or arbitrary, that is, without evidential grounds, for that would be highly subjective, a mere fancy. Nor is it hypothetical or conditional, for then it would not be genuine, since we would entertain it, as it were, with our fingers crossed. Rather does belief arise in us, as we have seen, because it is thrust upon us by the nature of the reality with which we are in experiential contact.[181]

Our conceptualities are not the application of a Dedekind 'cookie-cutter', because the dough is not amorphous, but contains a structure of its own, which may frustrate and surprise us.[182] Indeed only in this way can we account for paradigm shifts within natural science.

---

[179] See for instance Chomsky, 1968; also, his *Logical Structures of the Human Mind* (1955) and *Syntatic structures* (1957), Lennenberg & Feuer were Whorf's contemporary opponents.

[180] Torrance, 1984, p. 40.

[181] Torrance, 1980, p. 13; see also Torrance, 1984, p. 197; Torrance 1996a, p. 281; McGrath, 2002, p. 134.

[182] Putnam, 1987, p. 27.

Torrance's view finds support in Kerr,[183] Swinburne,[184] and Kirk.[185]

Whilst there is this *possibility* of an *actual* correlation between our mental concepts and the reality to which they refer, there is, for Torrance as we have seen, no question of a *necessary*[186] relation. One of the characteristic demands of scientific discipline therefore is that we resist any attempt to order such a relation in line with our own wishes and self-projections, preferring instead a scientific faithfulness to the reality of the object, as it makes itself known to us. According to Torrance, from Descartes, through Newton and Hume, to Kant,[187] the development of the 'modern mind'[188] has included an unwarranted concentration on the subjective pole, leading all too easily to the 'supermarket' voluntarism of Feyerabend, in which "categories ... are beyond the possibility of criticism or modification by empirical components."[189]

Kerr's 'grids' (quote in footnote 183), which our personal nature and mental make-up necessitates our use of, are too often surprisingly prescient of future developments and yet seemingly flexible enough to respond to corrections, than the voluntarist, who assumes they are nothing more than human constructions,[190] might predict. When colour words were added to language, the idea of colour as correlated in quite a linear fashion to wavelength was impossible to foresee, yet the latter quantity does seem remarkably well related to the former quality.

---

[183] "Things do not reveal their properties to us as if we were totally passive recipients, with no contribution of our own to make. Nor are we absolutely free to impose whatever grid we like upon the raw data of sensation. The colour and number systems belong in the realm of that interplay of nature and culture which is 'the natural history of human beings,'" Kerr, 1997, p. 104.

[184] "We believe our beliefs because we know that we do not choose them, but because (if we think about it) we believe that they are forced upon us by the outside world," Swinburne 2001, p. 40.

[185] "No sane person can maintain that it is always up to us what impinges on our sense organs ...

That being so, the suggestion must be merely that we *organize* the torrent of sensory stimulation which is largely independent of us. That is hard to dispute, but a long way from saying we construct reality." Kirk, 1999, p. 52.

[186] Colyer, 2001a, p. 324.

[187] Colyer, 2001a, p. 325ff.

[188] Torrance, 1984, p. 37.

[189] Colyer, 2001a, p. 330.

[190] Hacking, (1975, p. 27) quotes Foucault's example of fantasist Borges' supposed citation of a Chinese encyclopaedia in which animals are divided into: belonging to the Emperor, embalmed, tame, sucking pigs, sirens, fabulous, stray dogs, included in the present classification, frenzied, innumerable, drawn with a fine camelhair brush, *et cetera,* having just broken the water pitcher, that look from a long way off like flies.

Another example is the taxonomy of species. When biologists first categorized species into phylum, class, order, family, genus, etc. they had no idea of the objective relation between species and their DNA codification, and indeed DNA codification has shown up some mistaken relationships and associations, but on the whole, our first construal of the tree of living organisms was pretty accurate.[191]

'Grids' are fundamentally posterior to and contingent upon our engagement with that which we wish to categorize – vitally there can emerge good grids and bad grids, right grids and wrong grids – and such judgements are not made on the Feyerabendian grounds of what they do for us, but on the objective grounds of how well they respect the nature and inherent structure of reality, even if the relation between cognitive grid and inherent structure is as logically problematic as the analogous one between signifier and signified – there are intuitive and tacit components in this necessary integration of form. Therefore McGrath points out that such social construction should not be too cheaply equated as being tantamount to a rejection of objectivity: "objectivity and social construction are thus not *contradictory* (as both the Enlightenment and post modernity appear to believe)."[192] The question is again Hacking's: social construction – yes – but of what and how much?

If the enlightenment too easily contrasted objective knowledge with personal and socially constructed knowledge, then Feyerabend is equally guilty of the opposite, namely that once we have established the personal and socially constructed *elements* within 'objective' knowledge, we have successfully deconstructed its objective credentials. Torrance does not fall into this trap.

Personal objective knowledge cannot therefore be understood entirely independently and abstracted from its socio-linguistic setting, for the concepts and constructs it uses will be developed within that setting, and will contain corresponding implicit interconnections, associations and natural interpretations: hence the indeterminacy yet possibility of translation. Objective knowledge cannot be entirely supra-contextual or supra-linguistic, it is both a product and part of that context, and its unfolding exposition will therefore consist in the

[191] Torrance, 1969, p. 264. The opposing view is that of Couvalis (1989, p. 118ff) who argues that there are enough mis-categorisations (e.g. mice, cedars, jade, frogs,) for rebuttal.

[192] McGrath, 2002, Vol. 2, p. xiv.

hermeneutical task which dialectically seeks to describe knowledge with reference to context.[193] We note that that dialectic can never be frozen or made to stand still, it is an ever moving tradition.

Objective knowledge is tradition-mediated as per MacIntyre, not tradition-determined as per Feyerabend. Thus present even within mathematics and the natural sciences is a hermeneutical task of sympathetic imagination, as our own horizons are conjoined with those of other cultures and traditions.[194] Certainly, in the West our inheritance of the Greek tradition, so pervasive in certain sciences such as mathematics, implies that little imagination is required, but there are occasions (such as the transition from the concepts of impetus to momentum in physics) when the hermeneutical imagination has still to be exercised to adapt knowledge from past contexts. Feyerabend's worries concerning the comparability of incommensurable systems always assume that such systems are only ever in dialogues conducted within an arena of conflict – i.e. we are comparing in order to establish a superior. Therefore we need to 'defend' the weaker incommensurable theory from its stronger rival.

However rather than simply defending the weak – though we do not deny that there are times when that will be appropriate – a more constructive development of his incommensurability thesis would be to explore the possibilities for hermeneutical and dialogical operations, where *logical* ones will not suffice – whereby each tradition seeks to sympathetically integrate, appropriate and correct the other, so that dialogue and at times synthesis can be genuine – what, after all, Feyerabend says that he is after: though to be fair he is prescribing medicine – sedation perhaps – determined in part by the symptoms of his own time and context. If Western natural science is the chief offender, then we may need at times to fight fire with fire, but surely it is better to plan for peace by encouraging that tradition to participate in open dialogues. Feyerabend correctly identifies the need for open dialogues but seems to do little to promote them.

---

[193] McGrath, 2004, p. 110.

[194] For instance within mathematics, there are several techniques, often with different cultural backgrounds, which perform the elementary operation of multiplication – and it takes a fairly proficient mathematician to show that they amount to the same thing: e.g. write down two numbers you wish to multiply; underneath each number write two columns of numbers, halving one number and doubling the other, ignoring remainders; cross through both columns where the halving number is even; add up remaining numbers in doubling column. This is an ancient method of multiplication but to many educated in Europe and America today it must look as alien and unlikely as voodoo.

CHAPTER SEVEN

COHERENCE AND LANGUAGE

Epistemic coherence and correspondence

If theological science is to be able to speak realistically – albeit critically and convergently – of its object, then it must be possible for human language to refer – albeit analogically – to the divine being. As we have seen this is only achieved through grace (p. 109). Grace is gift, and our knowledge of God is therefore dependent upon us receiving that gift unhindered by pride or self-assertion.

> Irenaeus writes that unlimited though he is, God is become finite in his Son, for he has accommodated himself to our small measure lest our little minds should be overwhelmed by his immeasurable glory.[1]

However since this correspondence between statements and being is an act of grace, there can be no demonstration of it on other grounds – we can only, as Wittgenstein says, show rather than tell.[2]

Whilst all we may be able to demonstrate ('represent the logical form'[3]) therefore, according to Torrance, is coherence, that is not to reduce truth to coherence. Rather his realism implies that our theological statements are true in that they refer and have "correspondence with the objective reality of God's self-revelation in Jesus Christ,"[4] though we cannot represent the logical form of this relation, save fideistically and transcendentally. "We maintain, then, that theological statements are true in virtue of having a real term of reference beyond to which they point."[5]

Torrance rightly maintains that coherence is not value neutral when it comes to correspondence.[6] Certainly self-coherent fairy tales are

---

1 Torrance, 1994, p. 42.
2 Wittgenstein, 2001, 4.01f/4.12f, esp. 4.121; Torrance, 1969, p. 183.
3 Wittgenstein, 2001, 4.121.
4 Torrance, 1996, p. 1; see also: Torrance, 1969, p. 138.
5 Torrance, 1969, p. 182.
6 Torrance, 1969, p. 138.

always possible to construct, but their initial counterfactuals make them fairly easy to spot, so that more generally, coherence (as in a legal setting for instance) is taken to be a sign of, and a necessary condition for, the correspondence of statements to reality.[7] I believe that discussions about the relation of statements to reality *have* been unduly dogged by an inexplicable disjunction between correspondence and coherence[8] – in fact they nearly always point in the same direction, even if the former cannot be deduced from the latter, whereas often they are treated as rivals.[9] Feyerabend's theory of natural interpretations leads him to suspect that the counterfactuals of the natural scientific 'fairy tale' (his term[10]) are more subtle and less obvious, but this is surely Feyerabend at his most rhetorical. Very much more scientific and mathematical argument is based on the form 'if this were to be true, what other conditions might we expect to prevail', than one might expect by those who consider coherence the poor relation of correspondence.

Moore errantly describes the correspondence theory of truth as being "the view that there is in principle one true and complete description of reality,"[11] calling this position 'metaphysical realism'. The error is in the word 'one' – for there are correspondences that are well-defined, but are not one-to-one, or indeed 'onto', mappings.[12] What Moore is describing is a sub-category of correspondence: isomorphisms (see also previous reference to Polkinghorne: p. 161). All Moore's ensuing (and valid) warnings about the dangers of idolatry within (isomorphic) metaphysical realism[13] should not prejudice us against correspondence *per se*. At the end of the passage Moore cites Putnam as the source of this idea, a citation repeated by Patterson: "The world consists of some fixed totality of mind independent objects.

---

[7] Plantinga, 1998, p. 284.

[8] Hacking, 1975, p. 131.

[9] See McGrath, 2002, p. 52 – "An excellent example of this may be found in the writings of Laurence Bonjour, where we find a correspondence theory of truth which is held in tandem with a coherentist theory of justification." In fact there are also deductive proofs (in mathematics) constructed by transcendental contradictions with a rigour that shuts off any coherentist escape route.

[10] Feyerabend, AM, p. 298, also p. 52.

[11] Moore, 2003, p. 104.

[12] Though the caveat 'in principle' may just rescue the point, for we can always say that we are *in via*.

[13] Moore, 2003, p. 214.

There is exactly one true and complete description of the way the world is. Truth involves some sort of correspondence relation between words or thought-signs and external things or sets of things."[14] However Torrance's critical realism permits and requires a more mediated, contextualised, and traditioned correspondence between words and objects, offering referential power to terms without claiming for such descriptions that they are 'one, true and complete.' For such realism does not deny the varieties of human language, context and tradition, which form components within scientific descriptions, it merely denies that they are in any sense decisive. Indeed critical realism admits that it may be the case that no such ultimacy of description can be achieved except as analogous to the limit point[15] of a mathematical series which itself appears nowhere in the series, because all descriptions are context sensitive and 'language ridden' – language and context being constantly in flux. Therefore whilst critical realism will enjoin upon us a dialectical approach in which syntheses are constantly thrown afresh into the unending spiral of hermeneutical description and reading – *that* plurality is entirely different from the plurality of Feyerabend's relativism in which no such judgements can be made. This is an important point worth repeating: *Feyerabend insists that we make judgements on humanitarian grounds, because any others are likely to be misleading and open to abuse. The critical realist Torrance insists that though such judgements are difficult and dynamic by nature, we still have a duty to attempt them, and the scientific task of removing political and historical anomalies inherent within that attempt becomes more, rather than less, important.* Critical realism does include the non-Feyerabendian claim that the referential power of texts entails that descriptions can be compared even across traditions, and that some will fare so badly that they can simply be termed 'false', even by those from a different tradition.

Conversely the danger of Torrance's and Barth's theological realism is that, partly from fear of such relativism, theological commitments are given an undue degree of protective prevenience, less they should

[14] Putnam, *Reason, truth and history*, p. 49 quoted in Patterson, 1999, p. 12.

[15] Miller calls this a limit case as opposed to limit simpliciter, mathematicians tend simply to distinguish between limit points of finite and infinite series – see Griffith-Dickson, 2005, p. 73f.

be damaged in the hurly-burly dialectic of critical realism.[16] At this point the scientific and fideistic character of Torrance's theology rub uncomfortably against one another.

Achtemeier[17] too is concerned about attributing to Torrance a correspondence theory 'simplistically conceived.' This amounts simply to a warning against using an earlier Wittgensteinian theory of 'picture language,'[18] and Achtemeier points the readers towards the fifth chapter of Torrance's *Reality and scientific theology* where Torrance defends a 'deep and holistic' correspondence. Achtemeier's destination is a capable, and to my mind compelling, demonstration that, though they have different 'starting points', with regard to Torrance's realism and MacIntyre's tradition-mediated conception of rationality and justice, upon examination "the convergence is indeed remarkable"[19]: Torrance's 'social coefficient of knowledge' being equivalent to MacIntyre's 'dialectically-constituted tradition.'[20] Tradition for MacIntyre is 'the very form of rational inquiry', and therefore offers another account of tradition which closely matches Torrance's epistemology, and which also, whilst acknowledging the theory-laden nature of data, transcends the conflict model of traditions which fuels Feyerabend's radicalism, offering a progressive and constructive alternative.

The witness of the Bible corresponds to the Truth of God, known as personal being. Now the epistemic self-sufficiency of the 'strange new world within the Bible' limits how this correspondence can be established and justified, and so coherence within the Biblical witness – coherence understood as conformity to the person of Jesus Christ – will be the sign and demonstration of that correspondence. This reference is disregarded by Heidegger and Bultmann, on the grounds that this is not what the gospel is – the gospel is proclamation, not a set of propositional statements. However to posit these two options as mutually exclusive is another false dichotomy, and Thiselton is quite right to insist that forms such as address, promise and command, contain propositional content implicitly or explicitly.[21] Torrance (commenting on Calvin's hermeneutics) believes that Biblical statements have a

[16] Jeanrond, 1994, p. 175.
[17] 1994, p. 363.
[18] Waismann, 1965, p. 313.
[19] Achtemeier, 1994, p. 371.
[20] Achtemeier, 1994, p. 360.
[21] Thiselton, 1992, p. 294.

'*duplex sensus*' both of which must be understood for proper interpretation.[22]

Religious and scientific language

Torrance is, however, prone to fall into the opposite trap of assuming that theological language *only* functions in a propositional manner, and ignoring the mixed economy of theological speech: "the doings of religious language are legion."[23] Religious language does not simply play a scientific role, but an exhortative, emotive and rhetorical (in the Aristotelian sense) one.[24] So whilst one must resist the temptation to confine liturgical language to the expressive, declarative, directive, and commissive, denying it an assertative component, Torrance must also avoid the opposite temptation – both are false disjunctions.[25]

Torrance therefore offers us no account of the mechanism by which religious language can be constructed in a manner analogous to natural scientific language. The scientific usage of terms, aiming at refinement, as Achinstein[26] and Gadamer[27] point out, moves in the direction of univalence, where terms can be pinned down as 'well defined' – such as atomic number becoming isomorphic with the definition of elements, or DNA codings with the definition of species. It is notable that such univalence is not always immediately attained, and such refinement may develop from a process of empirical discoveries.[28] However this process may be legitimately artificially constructed, or at least given a helping hand, by scientists, so that, at least for a while, terms have scientific and non-scientific fields of reference ('peanuts' are scientifically speaking peas but not nuts, 'work' another example). However in contrast to the refinement of technical terms, in other contexts words may be 'stretched', by their use for instance in liturgy or poetry, so that the polyvalency of their reference and metaphorical potential increases. Torrance does not pay sufficient heed to this

[22] Torrance, 1988, p. 34; ultimately the distinction between *fides qua creditor* and *fides quae creditor* can be unhelpful.
[23] Blackburn in Moore & Scott (eds.), 2007, p. 56.
[24] See Avis, 1979, on the poetical (hymnody) and rhetorical (homiletics) use of theological language.
[25] Patterson, 1999, p. 50.
[26] Achinstein, 1968, p. 46.
[27] Gadamer, 1975, p. 375.
[28] See Dummett 1978, p. 428f.

polysemic and polymorphic 'stretching', preferring scientific refinement, and therefore gives insufficient consideration to how metaphors and models function within natural and theological science.[29]

[29] The Athanasian distinction (Torrance, 1988, p. 129; Athanasius, *De Decretis*, p. 157) between predicating a description of God and of a creaturely reality means that religious language will have no option but to employ language metaphorically (when the negative of the metaphor also has truth value) and analogically (when the negative has no truth value). Thus we may speak of God, according to Torrance, as positively analogically (as does Farrer, 1948, p. 47; Torrance, 1969, p. 294), but also balanced with due Barthian caution (CD I/1, p. 180). By the incarnation and the *lalia* of Jesus we alleviate much of the angst surrounding the possibility of this balancing act. For there God, often by way of παρα-βαλλειν, gracefully condescends to be known by human language and through human witness in order that we might speak accurately of God, not by reason of the Thomist *analogia entis* but by the *analogia fidei* (Barth, CD II/1, p. 227).

With regard to the relationship between natural and theological science therefore, we note that models and metaphors are employed, within natural science as much as theological science – indeed their use seems imbedded in the very nature of language (Soskice, 1985, p. 99; see also Peacocke, 2001, p. 15). Soskice's 'rough definition' of metaphor is "that figure of speech whereby we speak about one thing in terms which are seen to be suggestive of another" and model as a particular linguistic sort of metaphor (1985, p. 15).

Any scientific employment of metaphor will note with due care the limits of its applicability and the possibilities of inference, entailed within the polar relations of similarity / dissimilarity that emerge in that employment. So, for instance, in using the wave model of light, a decisive corner was turned when it was understood that the metaphor could be used without the implication of a propagating medium (ether). So models and metaphors are likely to trip one up, and carry over into the nature of the thing considered, incorrect but undetected artefacts from the sphere of dissimilarity into the sphere of similarity. "Theological analogies, it must be stressed, may mislead just as much as their scientific counterparts through misconstrual of their ontological implications, or a failure to respect their proper limits" (McGrath, 2003, Vol. 3, p. 125). However one cannot, prior to employing the metaphor, circumscribe the spheres of similarity and dissimilarity, for if that were possible then our need of metaphor and model would be short-lived. Our ignorance of the relation of the spheres, the fact that the similarities and dissimilarities are 'known unknowns', is precisely what gives models their heuristic potency. A metaphor without such imaginative possibilities is one which is still-born and useless, and risks dominating "the stage as to impede further discovery" (McKinney in McKinney (ed.) 1976, p. 44). So too there exists the corresponding likelihood that unexpected and unknown dissimilarities, as well as similarities, will arise in our employment of models, stopping us in our tracks, and forcing us to re-examine our assumptions and theoretical frameworks, thus reinforcing Torrance's conception of science as essentially realist (Soskice, 1985, p. 123). Therefore there exists, at times of revolution, a tipping point, at which one model gains a definite heuristic advantage over a previously superior one. In natural science we are able to investigate the possibilities of the model under the initiative of our own investigations, but in theological science we are restricted by how far God in his sovereign freedom, chooses to reveal himself to us. We have to live with uncertainty in our theological models in proportion to the mystery of God's being.

So models must not simply be understood as rungs on a ladder, which can be kicked away once some higher level of understanding is reached. As Torrance himself says,

Theological language will always be bound to carry its historical, symbolic and liturgical baggage,[30] its natural interpretations, along with it. Frege's distinction between sense and reference overshadowed his further distinction between *Sinn* and *Färbung* ('colouring'),[31] which he regarded as equally important for natural languages (of which therefore he was no fan at all).[32] Whilst it is possible, as Torrance wishes, to isolate artificially the scientific use of religious language within the academic study of theology, one must note the artificiality of that isolation ('laboratory' rather than 'real world' testing), the primacy of the liturgical context, its analogical[33] and sometimes metaphorical nature,[34] and that setting's refusal to be bound simply by considerations of scientific method.

---

doctrines are 'disclosure models' (Colyer 2001a, p. 363; Marley, 1992, p. 11, as opposed to 'picturing models'). Often the model and theory are so intertwined as probes of reality, that one is inseparable from the other: the model becomes the theory (Soskice, 1985, p. 115). In cases such as these, scientific models are to be understood as our ultimate expression of reality (Soskice, 1985, p. 107) and therefore for realist positive theologians such as Torrance, the economic model revealed by God of his essential self, is not a 'staging post picture' of God, but the ultimate statement about God, at least until God makes himself known in some new and decisive *act* of revelation. The properly termed 'model' of the Trinity (model ≠modal) cannot be dissolved or dissembled in either direction towards Unitarianism or Tritheism (Soskice, 1985, p. 112). As in natural science, ambiguity within the model cannot be prematurely deconstructed away without remainder, but must remain as part of the representation of reality (Soskice, 1985, p. 133).

The distinction between 'literal' and 'metaphorical' truth must therefore not be construed as a distinction between true and false, or otherwise relegate metaphor to a second rate or merely illustrative function (Soskice, 1985, p. 68) for statements which contain their metaphorical nature implicitly within them (Osborne, 2006, p. 124 – see also Young for the important distinction between compositional allegory and allegorical interpretation (in Watson (ed.), 1993, p. 112)). This view is the opposite of Van Fraassen, who takes empirical adequacy as the sole criteria of model validity, without any such realist pretensions regarding truth value: "My view that scientific activity is one of construction rather than discovery: construction of models that must be adequate to the phenomena, and not the discovery of truth concerning the unobservable" (1980, p. 5).

[30] Putnam in Hahn & Schlipp (eds.), 1998, p. 471.

[31] Dummett, 1978, p. 93.

[32] Frege, 1997, p. 303; Dummett, 1978, p. 116, – The distinction being that whilst sense and reference are distinguished by their relevance to truth conditions, sense and colouring are distinguished by the former having the property of being objectively communicable (Dummett, 1978, p. 424). Thus sense is part of 'thought' (which for Frege was "objectively communicable without residue" (Dummett, 1978, p. 442)) whereas 'colouring' belongs to the realm of emotions and feelings. Russell argued on the contrary that sense has no public meaning above reference (Hacking, 1975, p. 72). Frege's distinction then seems roughly equivalent to Aristotle's and Lacan's distinction between 'language' and 'voice' (see Griffith Dickson, 1996, p. 263).

[33] Barth, CD II/1, p. 227.

[34] See Soskice, 1985, esp. p. 68, 123.

Liturgy may use religious language in a way that for the theological scientist is tantamount to abuse, but the academy, Torrance would agree, can only act as fool and servant to the church, not its master. If liturgy contaminates language, theological science will have to reproduce and then factor in the contamination in the laboratory, it cannot eliminate it. With regard to theology Wittgenstein is correct and Frege wrong: theology cannot abolish the natural language of Christian life, it can only seek to understand and clarify. The construction of technical and scientific definitions of words, which correspondingly determines precisely their reference,[35] cannot be imposed in theology as it can be in the physical sciences. The fundamental activity of the church is worship, not scholarship – though Torrance is insistent that it is reference to Christ which is determinate for theological language,[36] and Barth admits that the only competent scientific authority is the church.[37] Cupitt here concurs: "'Religion is the test of theology and not *vice versa*.'"[38]

Torrance's blinkered focus on the 'scientific' function of religious language does leave him here unnecessarily vulnerable to Cupitt's non-realism. Key to Cupitt's approach is a disjunction between the doctrinal, ethical and spiritual[39] components of Christianity, supplemented by the unlikely contention that the eighteenth and nineteenth centuries' emphasis on the doctrinal was an historical aberration. This disjunction between worship and realism[40] is quite clear cut and operates as a chapter heading. Cupitt then asks what is in fact primary to Christian faith of these three, and not unreasonably comes to the conclusion that it is the spiritual. It is precisely at this point that Torrance, with his unitary approach should not be drawn into giving an answer, whereas his constant emphasis on the scientific means that there is a suspicion that he is implicitly falling into Cupitt's trap. In fact Torrance is quite at liberty to say that his concerns are with doctrine because he has a vocation as a systematic theologian, and that vocation is lived within the broader life of the Christian church – not all Christians are

35 Achinstein, 1968, p. 46.

36 Torrance, 1980, p. 116.

37 Barth, CD I/2, p. 756.

38 Cupitt, 1980, p. 38.

39 Cupitt, 1980, p. 99, see also p. 34f.

40 Cupitt, 1980, p. 56ff: "To go back to the distinction from which we began, he [Kierkegaard] is an expressivist rather than a realist," p. 66 – thus also applying a particular interpretation of Kierkegaard, at odds with Torrance's.

called to be systematic theologians. What Torrance needs to do to avoid the trap is nothing more than defend the view that there is a not insignificant doctrinal / propositional component to Christian language.

Revealingly Cupitt, in contrast to Torrance and Feyerabend, goes on to oppose religious and scientific language[41] precisely on the basis that the former is voluntary and polysemic, and that the latter is realist and univocal. Furthermore he refers to an 'archaic unity' which is lost – a unity which both Feyerabend and Torrance, in their own ways, are striving to recover – a unity of representation not wholly unrelated to the 'archaic' representations in art so enthusiastically endorsed by Feyerabend.[42] For Cupitt it is simply a matter of 'fact or value', religion or science, they are entirely disjunctive: "The choice must be made: it is one or the other."[43] Blackburn quite rightly rejects this: "I believe then that the expressivist theology cannot be true to the functions that religion centrally serves. There is no evading the fact that the person in the pew needs the ontological dimension."[44]

Coherence in theology and Scripture – witnesses to a single truth?

Barth and Torrance however, makes two claims for coherence which are not universally accepted. Firstly, with regard to the Biblical witness:

> The contents of the Bible constitute a single witness, and this is to be understood strictly and exclusively as the witness of God's revelation of grace in His covenant with Israel and in the fulfilment of the promise of the Messiah of Israel establishing this covenant, and therefore of the incarnation of the divine Word and the outpouring witness.[45]

Secondly, that the systemising of that witness within dogmatic theology, must aim towards a state of unified coherence.

So firstly whilst, as we shall see, Torrance is not making a claim for Biblical inerrancy, he is for the inner coherence of the Biblical material

[41] Cupitt, 1980, p. 122.
[42] Feyerabend, AM, p. 169ff.
[43] Cupitt, 1980, p. 126.
[44] Blackburn in Moore & Scott (eds.), 2007, p. 58.
[45] Barth, CD II/1, p. 108, also see p. 123, 208.

as witness. This is a disputed claim,[46] liable to lead to a 'canon within the canon.'[47] Furthermore there seem to be parts of the witness deliberately written or included in order to incorporate genuinely opposing views: Ruth & Jonah vs Ezra & Nehemiah for instance. As we have also mentioned, Jewish theologians would maintain that the Hebrew Scriptures, far from pointing unambiguously towards Jesus Christ, can be understood as perfectly coherent within themselves, without any reference to the books of the New Testament. The unitary witness of the Biblical material is something which Torrance must argue very hard for, if it is not just to be an assertion. In fact Barth[48] seems to be arguing for something stronger than that – not simply that the Biblical witness is coherently unified, but that it attains a role as a standard of coherent unity, by which other belief systems are judged. It is difficult to understand what this might mean in practice – it is one thing to say that below surface distortions and accretions a set of documents tells a unified story, it is quite another to say that that story is *the* 'plumb-line' or 'ruler' of unity.

Secondly, though one should not, like Feyerabend, argue from diversity to relativism, the 'core theories' of the paradigm of systematic theology, even if that is restricted to Reform, Catholic or Orthodox theology, seem slim compared to the up-for-grabs 'auxiliary theories'. Torrance is, for instance, here at odds with Lindbeck, whose purely coherentist approach identifies different denominational systematic frameworks as constituting the grammar wherein further work is carried out. Moore correctly identifies a defect in Lindbeck's approach as requiring a quite definite disjunction between theological statements as they are found in first order discourse (worship – where they have reference) and second order discourse (doctrine – where they need not).[49] Torrance's book *The ground and grammar of theology*, as the name implies, rightly supplements Lindbeck's grammar with the ground of a critical realist account of theological revelation. The historic particularity of the incarnation entails that our source of knowledge about that revelation will, for better or worse, be known by us through the traditions and history of particular people. Whilst it is

[46] "The Bible is an unruly book which, if we take it seriously, refuses to be locked up in any systematic cage." Blank in Küng and Tracy, 1989, p. 261.
[47] Osborne, 2006, p. 360.
[48] See McDonald, 2000, p. 224.
[49] Moore, 2003, p. 98.

evangelically advantageous that that knowledge is also then re-framed as it is introduced to other traditions, we note that the transmitter tradition will also exert a reforming influence over the receiver, as core Hebraic patterns of thought informed the Hellenistic when the gospel was first made known to the nations.

At the systematic level, to employ Kuhnian terms, the danger for Torrance is that scientific theology has not reached a normal phase, where basic contexts and frameworks are held in common, providing a stable platform for more detailed work. Whilst Torrance is consistently critical of those who coerce Biblical theology into an alien framework, arguably this is the only way that the Bible can be made to speak univocally, and his own 'distillations' and attempts at 'integration of the forms' may be viewed as having some of the same outcomes if not the same aims – critics may contend that good scientific method demands that we shall have to leave Biblical theology in a less determinate and more fragmentary state.

For example, whilst Christian theologians are almost by definition liable in some form or another to assert the 'supremacy of Christology in our knowledge of God', for Barth and Torrance that supremacy is radicalised, so that a Biblically known Christology is *contrasted* (or at least becomes the yardstick to which others must measure up) with and relativises other possible sources of theological knowledge – sources which might more naturally be supposed to be complementary: natural theology, knowledge of God as encountered in the Hebrew Scriptures, pneumatologically informed knowledge, either by charisms given to particular persons or offices, or through the corporate life of God's church. This prematurely forces systemic theology towards normality, on the face of it by means of denominational prejudice.

In another example, considering Athanasius' work on the doctrine of Christ, Torrance must contend that though Athanasius introduces extra-Biblical terms such as *homoousion,*[50] Athanasius does so not with a speculative spirit, as though the Trinitarian picture were one framework amongst several for systemising the Biblical data into Christian dogma, but by careful analysis of that data, seeking, uniquely and utterly faithfully to it, to represent both how God really is towards us in Jesus Christ, and also how he is for himself in eternity.[51] The opposing

[50] Torrance, 1965, p. 40; Purves in Colyer, 2001.
[51] Torrance, 1984, p. 255f.

Arian view is not simply another choice, but an inadequate one, given the facts of the situation.[52] In the same way Hooker seeks to show that the path to Chalcedon was not just a matter of choosing one path from many, but was 'an inevitable development.'[53] However whether that inevitability was ideational rather than political is, as we have seen, more problematic to maintain, noting also that this interpretation of Athanasius is contentious.

The alternative Feyerabendian view would be that theoretical pluralism should be a feature of dogmatic theology. Why assume with Torrance that we are at least aiming towards one dogmatic framework which is internally coherent and systematic, and simultaneously faithful to the Biblical text? Theoretical pluralism would have profound and perhaps advantageous implications, not least within the ecumenical world. If it were proposed that there were various such dogmatic frameworks, Roman Catholic, Orthodox, Reformed, Anglican, Lutheran etc., which were different, and yet fared equally well on internal coherence and faithfulness to the Biblical text,[54] that would significantly modify what we might expect from ecumenical discussion. Thus Lindbeck argues that the only criteria by which we can assess dogmatics by is how that theory "organizes the data of Scripture and tradition [is] with a view to their use in Christian life and worship," rather than any external reference they might have to the being of God[55] – a instrumentalist view very similar to Feyerabend's insistence that human flourishing must be the ultimate criterion of good science. As Feyerabend would also point out, just as within natural science, the fact of the matter is that such diverse structures do exist, and on-going disagreements over what merits a properly 'Biblical' theology, but this is neither an argument for or against theoretical monism or pluralism.

We also note with Feyerabend, that just as with natural science, it is a fact that arguments amongst Christians over whose dogmas are right, has not been terribly fruitful and has produced a great deal of pain and suffering. If it is true that denominational differences are incommensurable, and that formulations of dogma are in a good measure *chosen*,

[52] Torrance, 1965, p. 40; Torrance, 1996, p. 12.

[53] McGrath, 2006, p. 156; see also Polkinghorne, 2007, p. 56; Colyer, 2001a, p. 72.

[54] Indeed most discussion of which form of systematic theology is the most 'Biblical', does not revolve around direct comparison of texts, but in trying to establish what such faithfulness consists of.

[55] Lindbeck, 1974, p. 106.

being roughly equally faithful to the data, then ecumenical dialogue has an entirely different dynamic from a conception, which includes, beyond the clarification of dogma and barriers to dialogue, not only 'agreed common statements' of doctrine between churches, but finally convergence to full ecumenical doctrinal consensus. However if Patterson is right to identify a 'key difference' between 'realist theology' which asserts that "Christian truth claims ... correspond to an extra-linguistic reality," and Lindbeckian 'postliberal theology' which "views truth in terms of faithfulness to the norm of Scripture as interpreted by tradition,"[56] then one wonders firstly how any clear water within that difference is reconciled with the realist Barthian 'bondage to Scripture', and secondly, where the moral seriousness of the postliberal tradition lies if its objective omits any reference to the extra-linguistic significance of Scripture over any other literature, which too must be understood within its context and tradition. If we are not even aiming at corresponding Biblical and theological statements with an extra-linguistic reality, then why bother with them over any other writings? As literature they are by no means without peers, but surely what marks them as Scriptures is rooted in that correspondence. Patterson goes on to admit,[57] correctly in my opinion, that the dichotomy is in fact a difference of emphasis, between Putnam's internal and external realism, which rests "on assumptions that indicate the inadequacy of present models."

We can imagine Feyerabend the theologian, gleefully dreaming up new doctrinal frameworks which 'pleased him for one reason or another', holding them with great tenacity at the World Council of Churches, perhaps even founding new denominations, on the basis that such frameworks are just as faithful to the Bible as those of Thomas Aquinas, Martin Luther, or Karl Barth. After all, McGrath points out, Augustine maintains that some matters of doctrinal interpretation can only be decided by factors beyond those of Biblical witness and internal coherence.[58] However given likely historical assumptions concerning the stability of the programme's core, what Augustine is envisaging is more akin to Lakatos research programme than Feyerabendian pluralism – core doctrines surrounded by a protective belt of contingent auxiliary interpretations.

[56] Patterson, 1999, p. 3.
[57] 1999, p. 4–5.
[58] McGrath, 2001, Vol. 1, p. 61.

Indeed it is a model which is familiar within both ecumenical and intra-denominational discussions – certain 'fixed' standards: the rule of scripture, the rule of the creeds, the Lordship of Christ, the doctrine of the Trinity, the statements of certain sets of ecumenical councils, the teaching of the Magesterium, a certain confession (Westminster, Belgiac, Scots, etc.), – held together with diversity and flexibility on matters not made unambiguously explicit within those standards and with regard to the interpretation of those standards themselves. Torrance, who was very active in ecumenical discussions,[59] believed that our search is simply for *one* systematic theology, not to be completed this side of the *parousia* no doubt, but a single point of convergence implied by the simplicity and perfectedness of God himself. That is the aim of a realist scientific theology, and was also the aim of early Feyerabend's super-realist conception of natural science.

Within the community of the Christian confession, the coherence of that confession is not self-evident. Many people upon a fairly brief encounter with Christian doctrine find it to be rather incoherent, divided and confused, though within post-modernity, tipped off by Feyerabend, many are becoming more aware of similar divisions, confusions and complexities within natural science.[60]

Scientific theology is the labour in which we work hard at uncovering a coherent structure to God's revelation, analogous to the way in which the theoretical physicist has to work hard to uncover the coherent structures of a reality, which on the phenomenological surface seems wildly un-homogeneous – to the layman relativity theory is even more vexatious than theology, though he may place his trust in a more univocal community of experts. Feyerabend would warn him against such trust in a faction with its own aims, commitments, and products, pointing out the times when there has been a far more univocal theological community. Presently however the lack of agreed results and the proliferation of 'theologies', both academic and denominational, whilst it may please the Feyerabendian, is a concern for academic theology, though again one must not be tempted by an argument from diversity. Certainly it may 'settle down' – after all there have been

[59] Colyer, 2001a, p. 46.
[60] Osborne, 2007, p. 349.

periods of 'normal' theological science, between its 'revolutions'[61] – but it shows no sign of doing so, and despite progress on the institutional ecumenical front, increasing theological polarisation now tends simply to transcend denominational boundaries – witness the current state of the Anglican communion.

[61] It is possible to think of theological science in such Kuhnian terms, with schisms such as the Reformation constituting a scientific 'revolution' – see Küng and Tracy, 1989.

CHAPTER EIGHT

FROM FOUNDATIONS TO SPIRALS

The final stage of the puzzle will therefore be to answer the following question: if as Feyerabend says, knowledge is now 'without foundations', does that therefore also entail the epistemic anarchy which Feyerabend advocates, or is it possible to salvage the rather more ordered construction which is theological science?

Enlightenment foundationalism and reductionism

Logical positivism represented the most systematic formulation of enlightenment (and empiricist) foundationalism,[1] and Feyerabend's correct critique (theory ladenness/determinedness of observations) of this particular form of foundationalism is over-stretched to rather shaky conclusions (theoretical pluralism, counter-inductivism, relativism, voluntarism), not shared by other more moderate critics of logical positivism. Torrance for instance, like most theologians, rejects logical positivism, welcoming a renewed 'dynamic integration'[2] between the theoretic and empirical components of modern physics.

It is not obvious, at least from a developmental psychological or cognitive point of view, why foundationalism provides an ideal metaphor for knowledge or the methods of its acquisition. The two features of foundationalism which were attractive to enlightenment thought were

---

[1] See Oberdan in *The British Journal for the Philosophy of Science,* "The Vienna Circle's 'anti-foundationalism'". Vol. 49/2, p. 297–308, for a 'traditional' account of Schlick's work as foundational against the 'anti-foundational' interpretation of Uebel. For another anti-foundationalist account of logical positivism see M. Friedman, *Reconsidering logical positivism*, CUP, Cambridge, 1999. Here we have adopted a more conventional account of logical positivism, firstly because it is beyond the scope of this book to weigh arguments within the philosophy of science regarding traditional and revisionist accounts of logical positivism – but more importantly because our major players Feyerabend, in particular, Torrance, McGrath (e.g. 2002, p. 32, Pannenberg: "empiricist positivism take experience of more accurately, sense perception, as the ultimate fact upon which all knowledge builds," 1976, p. 29) are all working with a 'traditional' foundationalist understanding of logical positivism.

[2] Torrance, 2001, p. 77.

the possibilities of progressiveness and reductiveness – the degree to which these qualities were present in any particular formulation of foundationalism was a function of the degree of separation between the floors or levels of the foundational structure. Thus logical positivism promised a high degree of both, since it sought to place a hermetic seal between the levels of observation and theoretical, to the extent that the reductiveness of the system was in fact liable to collapse the building by denying the theoretic level any ontological purchase. The risk was that in terms of ontology, natural science was in danger of being reduced to a very long list of observations. If the upper layers of the superstructure can be reduced to merely an orderly summation of what lies in the empirical and self-evident foundations, then they begin to look ontically superfluous.

The failure of higher levels to attain ontological reference means that they subsist at the level of inductive or deductive corollaries of the foundation. However if they are granted a degree of ontological autonomy which transcends reduction, therefore permitting a degree of osmotic interpretive activity between the levels, then the orderliness and progressiveness of the levels is threatened, as we can no longer build assuming the soundness of our foundations. That this represents a more faithful account of how natural scientists actually behave has not gone unnoticed[3] – they are constantly going back to the data, re-examining and scrutinising it, developing sometimes rather speculative extensions to the theory which in turn are submitted to empirical testing. It is therefore too simplistic and not true to life to describe natural science as a series of theories built upon observational data.

Not only, however, does this form of foundationalism seek to offer an account of the structure of natural science as being a set of theories build upon self-evidency and sense-data / empirical observation, but it also must account for the variety of objects and concepts which are sometimes shared and sometimes not between different sciences. So for instance physics talks of particles and fields, chemistry of molecules and reactions, and biology of cells and organisms – yet all three interrelate. A pressing question for ontology is – whilst each discipline might utilize findings and terms from the other, and therefore respect the independence and competency of the other to rule on such entities – are the possibilities and actualities of higher level objects

[3] Ritchie, 1948, p. 22; Feyerabend, FTR, p. 190.

each ultimately capable of Laplacean reduction to the behaviour of elementary ones such as particles? Or does each science deal with entities which whilst bounded in terms of possibilities by the properties of lower entities, nonetheless possess the capability of emergent patterns of behaviour which are not dictated or determined by those lower orders, but better understood as lying within a web of mutual entailment and possibility? There is also the question as to how far concepts used in one discipline are then used properly and appropriately in another. The task of natural science can no longer be thought of as the depiction of reality as a long series of idealised objects in a field, but as an unfolding many-body problem, incapable of precise depiction, as constantly each emergent property sends consequential ripples up and down the overall structure.

Torrance adopts from Polanyi[4] a conception of the stratified nature of reality, described by the variety of sciences, including the natural and theological, which resists any such reduction, and grants ontological *gravitas* to the objects of each particular science. Rather than diachronically or logically attempting to build upon the foundation of lower floors, these strata are synchronic, and parallel investigations into the object of their sciences, whilst they are certainly constantly mutually interactive – for Torrance's holism implies that no object is entirely independent of any other object or class of objects – they nonetheless retain their own reference and domain of applicability. Their conceptual and methodological apparatus is constructed in response to an encounter with the object in question, and therefore each is fallible, not primarily by corrections in other layers, but in faithfulness to that on-going encounter. However secondary encounters within surrounding strata are illuminative and implicative within the ontic layer in focus, in setting boundary conditions and limiting possibilities. Thus Biblical studies can shed light upon and set limits as to how systematic theology can employ Biblical texts, just as it in turn has to take note of textual and philological findings, but it cannot demand on the grounds of an individual interpretative point that systematic theology construct itself in a certain way. There are crucial texts as there are crucial experiments, but here too under-determination entails that this identification as 'crucial' can only be understood from within a hermeneutical circle.

[4] McGrath, 1999, p. 233.

According to Polanyi and Torrance this is analogous to what happens in natural science, where perhaps the greatest reductive pressure is in the field of biology.[5] The lower levels of science control what is *possible* at higher levels, and give boundary conditions as to what might emerge, but they do not predict what in fact does emerge, and neither can they.

> Structures of thought co-ordinated in this way are necessarily open structures, always incomplete in themselves, but orientated beyond themselves; the different levels in which these structures are found to constitute a hierarchy, as Bertrand Russell has shown, which is open upwards and not downwards, and that statements are normally tested and controlled from two coordinated levels, but because the structures are open upwards and not downwards, there can be no reductionism.[6]

There is an insuperable deficit in the knowledge of the 'all-knowing' Laplacean demon which prevents it being deductively worked up to describe the abundant reality of the world as it actually is.[7]

In practice however, this sort of reductionism is ruled out not by purely ontological or epistemic principles (though at the quantum level that may be the case), but for practical ones of calculability. We do not, for instance, treat the human body as a collection of molecules and atoms simply because we have no way of knowing, recording and thenceforward calculating their type, position and velocity. Advances in bio-nanotechnology, which rely on the manipulation of crucial molecules,[8] may point towards that possibility, but we are a long way from telling how far in principle that process is possible. Polanyi and Torrance are in danger of drawing epistemic conclusions from practical limitations, even if the latter in themselves seem perennially insurmountable. Ultimately such conclusions *are* offered as a result of quantum and Gödelian theories but these results apply to such narrow sets of circumstances, that one should be aware of drawing too global conclusions. That Torrance wishes to do so is conditioned by the theological requirement that God, whose calculating ability is presumably

[5] Torrance, 1969a, p. 84; see also Popora in Archer et al. (eds.), 1988, p. 350.

[6] Torrance, 1984, p. 323; see also Torrance, 1996, p. 86.

[7] Polanyi, 1998, p. 140; 1969, p. 154; with Prosch, 1977, p. 29.

[8] E.g. Lu Jun-Hong; Li Hai-Kuo; An Hong-Jie; Wang Guo-Hua; Wang Ying; Li Min-Qian; Zhang Yi; Li Bin; Hu Jun, 'Nano-manipulation of single DNA molecules based on atomic force microscopy' in *Engineering in Medicine and Biology Society, 2005.* p.7478–81, 2005.

infinite, should not be bound by such determinism. Furthermore more recent developments in this field, particularly with regards to emergence[9] are not well understood within the natural scientific community and therefore are perhaps too easily seized upon by theologians (who understand them even less well).

Developments in chaos theory point in what for theologians may be undesirable directions – towards determinism.[10] For there is nothing indeterminate about chaotic systems, rather tiny differences in initial conditions are amplified into highly non-linear but fully determined states.[11] However Torrance and Polanyi are correct in supposing that it is difficult to see *in principle* how features such as consciousness and moral freedom might emerge from an account of the activities of elementary particles, however detailed.[12]

Feyerabend too, in reaction to his own earlier Popperian and positivist positions, is an anti-reductionist, but to more radical ends.[13] Feyerabend's incommensurability thesis implies (even diachronically within a single science) that the 'logical disjunction' between paradigms rules out reductionism from one scientific paradigm to another. If we are to stick with the realism which demands comparability, then at a global level there must at least be a contingent attempt at ontological reconciliation – even if we are prepared to put up with a little local instrumentalism on the way. However if we forego that realism, then each paradigm in terms of ontology is reducible, and ultimately *trivially so*. If we retain realism, then we need to give an account of the basis on, and (epistemic – Feyerabend's point is that we should

[9] See Smedes, 2004, p. 91 – 'mystifying' is his term.

[10] For an account of Einstein's determinism, and his argument with Popper, see Popper, 1969, p. 61.

[11] Polkinghorne (1998, p. 53f) attempts to make the contrast on other grounds, 'intrinsic unpredictability' versus 'mechanical' systems, rather than determinate and indeterminate systems. Practically at either end of the scale such a distinction works, however it depends upon a scale of rising complexity which has no particular dividing line. This is an epistemological rather than ontological distinction (also things do not have to be chaotic to be unpredictable – people for instance), and he admits that "there is no logically inevitable way to proceed from epistemology to ontology," so that critical realism is the philosophical position that respects and holds these realities in tension.

[12] Polanyi, 1969, p. 42; Wigner in Polanyi, 1961, p. 232; Swinburne, 1991, p. 161; Penrose, 1989, p. xv, 292.

[13] "The 'disunity of science' that is the central theme in Feyerabend's later work … anti-reductionism was, of course, a long standing theme in Feyerabend's philosophy of science …[quoting Feyerabend] 'science does not contain one epistemology, it contains many.'" – Preston in Preston, Munevar and Lamb (eds.), 2000, p. 81.

substitute these with moral choices) criteria by which, we are attempting the reconciliation, as well as accounting for theoretical and paradigmatic proliferation, even if this cannot be brought to systematic logical articulation. Torrance declines to offer Feyerabend this logical articulation, but nonetheless holds onto other criteria such as objectivity. Feyerabend rejects such a defence, but only by demolishing a straw man objectivism, and thus Torrance is able to avoid Feyerabend's deconstructive non-realism.

There are many sciences for Torrance, because within the created inherent structures of reality there are necessarily irreducible strata, but nonetheless there is unity in that these strata cohere into one reality, and that reality is rationally apprehendable and finds its ultimate source of being in the God who created all things.[14] By contrast in our voluntarist interpretation of Feyerabend the multiplicity of the sciences is ultimately an expression of the variety of human temperaments and traditions, and any search for unity, is pointless.

McGrath[15] too, drawing on Nicolai Hartmann, emphasizes that any realism which is flexible and critical enough to contain distinct ontic strata rules out reductionism. Critical realism is therefore a stronger rather than a weaker form of realism than naïve realism, because the latter restricts in advance the nature of reality and our depiction of it. Critical realism, by its revisability and contingency, actually allows reality continually to reveal its true form and shape, and is fluid enough to reject preconceived conceptions of what reality looks like, or how it can be represented, however counter-intuitive they turn out to be.

One final notable feature of this layered structure is that according to Torrance, and *pace* logical positivism, as one moves up the strata, "from the more tangible to the more intangible, we penetrate to things which are *increasingly real and full of meaning*."[16]

Progressive Foundationalism and Ordered Strata

Whilst enlightenment foundationalism is therefore forced to admit that natural science, being an historically unfolding phenomena in which fresh discoveries, observation and theoretical approaches are

[14] See again Torrance, 1985, p. 24.
[15] McGrath, 2006, Ch 5, opening section.
[16] Torrance, 1984, p. 195; Colyer, 2001a, p. 181.

constantly emerging, cannot be described in static positivist strata, it may still be possible to rescue an idealised 'snap-shot' foundationalism as a limit point of natural science, as an unrealisable goal, yet exerting a gravitational pull on the shape of natural science in its unfolding historical actuality. Feyerabend rejects this defence because theory and observation are in such a state of entangled tension that neither can pull clear enough to realise this pattern even locally, however thinly we slice up reality, so that it is an undesirable as well as unrealisable aim. Rather each sometimes overlapping slice must be analysed within alternative incommensurable frameworks, each with their own methodologies, conventions, and conceptualities. That some of these disciplines might band together and predicate the resulting integral 'scientific', is for Feyerabend nothing more than tactical move by a group of activities, sponsored and participated in by a particular sociological tradition, with the aim of furthering certain political and cultural vested interests held in common, by weight of numbers, and with a degree of imperialistic self-interest and protectionism.

Contrastingly Torrance would maintain that the strata are ordered to the extent that progressive verisimilitude is a possibility – though not always or necessarily a historical actuality – so that each stratum (and therefore the science of that stratum) retains both a degree of bounded dependence tensioned with a degree of elastic ontological freedom and autonomy. However Torrance would deny that the evidence of the senses and self-evidency are the ultimate ground of those sciences – notwithstanding the genuine utility of empirical data and deductive reasoning within many of them. The permeability and interdependence between the strata is strong but not ultimately incommensurable, so that the relations between strata still possess an orderliness and rationality (in the widest sense of the word) which maintains a sense of structured reality. Whether such a modified view of progressive foundationalism amounts to any sort of foundationalism at all, determines whether one regards Barth and Torrance as foundationalists or not. Hence the continued debate as to whether Barth should be regarded as a pre-modernist, a foundationalist, or a post-modernist – a debate which has achieved no firm conclusions. As Schwöbel puts it:

> One should not too easily see Barth as a post-modern theologian, perhaps even the first post-modern theologian. There is no doubt that Barth would have agreed with much of postmodernism's criticism of modernity, especially with the critique of foundationalism, … however it remains doubtful whether Barth would have found it easy to come

> to terms with the plurality of paradigms, the celebration of difference, sometimes the sheer arbitrariness of positions that seem to be by-products of the self-congratulatory attitudes of post-modernism.[17]

We speak of 'ordered strata' in contrast to Feyerabend, using the term in much the same way as Hacking's 'inherent structuredness'.

Fluid axioms

If the object of theological science is to be God, and the Biblical material its data, then it is clear that the positivist schema – observational data which commands a high degree of consensus, and metaphysics-'lite' theoretical terms which are verifiable by but do not contaminate observational terms – is wholly inappropriate. For not only is the observational data theory-laden, but it must be, in order to be theologically useful. Similarly the doctrine of God can hardly be considered simply an explanatory device devoid of metaphysical content.

Feyerabend's criticism of this positivist schema provides theological science with a defence against its implications, but his further radicalism is in danger of reducing all sciences, theology included, into his voluntarist relativist quagmire with few epistemic pretensions – Torrance's theological science, with its robust scientific realism, claims deeper ontic bite than simply being a tradition conducive (or indeed perhaps not) to human flourishing. Feyerabend's radicalism is manifested in that once the seal between levels is broken, then no clear relations between ontic strata can be drawn, and the effects of adjustments made in any level for any other incommensurable level are always open to voluntarist negotiation. So voodoo need not by necessity be affected by *any* of the results of Western natural medicine, but it may choose on pragmatic grounds to incorporate some of its findings tactically, if the lot of its practitioners and patients is bettered.

Torrance cannot permit such epistemic anarchy in an ordered and intelligible creation, and therefore must restore some structure. Given that the old foundationalism is not an option and his own Christocentric foundationalism is hampered by the epistemic mediacy of its data, he needs therefore to devise a system whereby the lower strata of theological science provide fallible and revisable 'axioms' for higher levels.

---

[17] Schwöbel in Webster (ed.), 2000, p. 33; see also McGrath, 2006, p. xii.

Torrance calls this working with 'fluid axioms,'[18] and preceding "according to the nature of the object" *(kata physin):*[19]

> The kind of scientific theology that arises in this way may be called *fluid dogmatics*. This is not the kind of theology that is developed through logical deduction from fixed premisses or axioms, taken from Holy Scripture, or from conciliar or ecclesiastical pronouncements, the typical theology of Roman or Protestant scholasticism, in which statements of the truth tend to be identified in a nominalistic way with the truths they are held to express. Rather, it is the kind of theology that develops under the compelling claims of the Word and Truth of God's self-revelation.[20]

Thus according to Torrance, a feature of scientific method is the willingness to hold[21] everything contingently (concepts, language, axioms, systematic frameworks, personal prejudices, experimental methods, research criteria) to be confirmed or repudiated by the encounter with the reality of the object itself – this feature of scientific method is the epistemic equivalent of theological and moral fallibility, together with the virtue of humility. This posterior method, according to Barth (*nachdenken* is his term) is typical of an older Protestantism (pre-1700) before the Kantian emphasis on the *a priori* made prior and distorting extraneous demands on Christian revelation and developed into the scholasticism which Barth sought to recover Protestant theology from.[22]

This method Torrance traces to Athanasius:

> The significance of the Athanasian method can be indicated by comparing it to the way in which we operate today, not from fixed axioms from which we proceed by logico-deductive processes to conclusions, but with 'fluid axioms' in field-physics as progressively refined conceptual instruments by means of which we seek to lay bare the inner intelligible structure of the space-time metrical field in some area of scientific investigation. It was as far as I am aware, Athanasius who first began to develop this kind of scientific method operating with relational concepts in his attempt to think faithfully and worthily of God in accordance with its

[18] Torrance, 1995, p. 227.
[19] Torrance, 1984, p. 312.
[20] Torrance, 1982, p. 49.
[21] 'Held' is just as important as 'contingently' – the problem for Feyerabend is that his radical contingency requires us to keep all the plates of fallibilism spinning all the time.
[22] Barth, CD I/2, p. 4.

> nature revealed in Christ and in accordance with the nature of his economic activity in our human and historical existence.[23]

We note that not only is this method appropriate for theological science, but it enables natural science to escape the axiomatically embedded 'damaging dualisms' between the sensible and intelligible,[24] which was only to be achieved in relativity and quantum theory,[25] and to live in structural conformity with the post-Gödel world.[26] Thus Torrance asserts that these dualisms have "caused deep splits in our culture with consequent widespread disintegration of form in the arts and social sciences."[27] Athanasius is not the only one credited by Torrance for his startling prescience of Einstein's work: "Philoponus offered an account of time and space in terms of the movement of light, which was an astonishing anticipation of relativity theory"[28] (though Torrance fails to note that on the down side he also rejected atomic theory[29]).

This 'Athanasian method' is in contrast to Augustine's incorporation of Platonic dualisms within his theology, which according to Torrance "frequently obstructed the advance of science"[30] – note 'science' not 'theological science'. Furthermore according to Hing Kau,[31] derived from this deep Augustinian Platonic cleavage between the 'sensible' and 'intelligible', was Augustine's distinction between 'scientia' and 'sapientia', with theology categorized as the latter. It is notable that this assessment is as much to do with Augustine's impact on science, as Torrance's estimation of him as a theologian – Torrance's views on Augustine are in fact more or less word for word those of Michael Polanyi: "St Augustine denied the value of natural science which contributed nothing to the pursuit of salvation. His ban destroyed interest in science all over Europe for a thousand years."[32] For these and other

---

23 Torrance, 1995, p. 227.
24 Torrance, 1969a, p. 15.
25 "The new science undermines epistemological and cosmological dualism," Colyer, 2001a, p. 204.
26 Hence the importance of the epistemic 'climate change' detected by Torrance and noted in chapter three.
27 Torrance, 1995, p. 228; see also: Torrance, 1969a, p. 15.
28 Torrance, 1996a, p. 274.
29 Philoponus, 1999, p. 65.
30 Torrance 1969, p. 58.
31 Hing Kau, 1996, p. 82.
32 Polanyi, 1998, p. 141.

reasons we find Torrance replacing Augustine with Athanasius in his roll-call of theological heroes.[33]

Conversely his assessment of Athanasius is by no means an uncontroversial one. Morrison[34] and Spjuth cite R. P. C. Hanson's challenge to Torrance's interpretation of Athanasius, as Luoma quotes a similar one of Meijering.[35] Hanson emphasizes the neo-Platonic elements within Athanasius' own writings, perhaps even more profound than those in Augustine, resulting according to Hanson in a docetic 'space-suit Christology'. Spjuth[36] furthermore argues that despite Torrance's explicit rejection of dualism, a 'remaining dualism' is therefore carried over from the theology of Athanasius into Torrance.

Certainly some passages (*On the incarnation*, p. 28) give rise to that impression, but only if, as Torrance maintains is not the case, one understands Athanasius as working with a Platonic 'container' view of space-time. If he has succeeded in transcending that limitation, as Torrance maintains, then the charge of docetism evaporates. Torrance is quite clear, more generally, that what occurred in the Patristic period was that Hellenistic patterns of thought were themselves being restructured in order that they might truly express the gospel to the Gentile world, even if the actual language was now Greek rather than Hebrew, so that lingering Greek natural interpretations were being made impotent in their utilisation in communicating a conceptually Hebraic gospel:

> It was impossible for Barth to agree with Harnack that the Chalcedonian formulation of the doctrine of Christ was an Hellenisation of the Gospel through the irruption of Greek and in particular of Aristotelian philosophical concepts into the understanding of the church. While important Greek terms (*ousia, hypostasis, physis, energeia,* and the like) were taken over, their meaning was so radically altered that their new use actually helped to transform the very foundations of Greek thought and culture.[37]

What is most notable about the 'Athanasian' method adopted by Torrance is his clear implication that such a method is appropriate (though in modified form) for natural science. The old foundationalism

[33] McGrath, 1999, p. 138.
[34] 1997, p. 288.
[35] Luoma, 2002, p. 89.
[36] Spjuth, 1995, p. 218.
[37] Torrance, 1990, p. 198.

is for Torrance not simply bad news for theological science, but bad news for science. Thus theological heroes are often also likely to be natural scientific heroes, and vice versa, so deeply interrelated are the theological and natural scientific contexts – if the 'climate' determines or at least influences the content of the particular sciences, then the 'climate setters' may come from any discipline. Indeed if one were to draw up a list of Torrance's scientific heroes and villains, we end with two rather unusual lists: Heroes: Athanasius, Philoponus, Calvin, Maxwell, Einstein. Villains: Plato, Aristotle, Augustine, Descartes, Newton, Bohr.[38]

> In spite of the unfortunate tension that has so often cropped up between the advance of scientific theories and traditional habits of thought in the Church, theology can still claim to have mothered throughout long centuries the basic beliefs and impulses which have given rise especially to modern empirical science, if only through its unflagging faith in the reliability of God, the creator and in the ultimate intelligibility of His creation.[39]

Torrance (obviously denominational allegiance is an underlying factor here) argues that whilst Luther inherits strong Platonic influences from Augustine, Calvin, though he often quotes from (anti-Pelagian) Augustine, looks more to the Eastern Fathers, especially Athanasius.[40] Calvin consequently, according to Torrance, *not only* initiates 'modern theology,' but 'modern scientific thinking'[41] – this is an ambitious claim indeed, and again the question is: according to which criterion is the importance of Calvin's work being assessed? It is worth another quotation to underline how unambiguously Torrance derives the development of modern natural science from the Reformation.

> It was this transfer of the centre or locus of authority from the subject to the object from the collective subjectivity of the Church to the objectivity of the one Word of God, that gave rise to the scientific method that passed from the Reformation into the sciences. Hence as a result of this change in theological inquiry there were written, not long after the final editions of the *Institutes,* works on scientific method by men devoted to the interpretation of the books of divine revelation (e.g. Andreas

[38] Torrance, 1969, p. 92.

[39] Torrance, 1969, p. 57.

[40] Torrance, 1988a, p. 82, Torrance accuses Barth of getting this wrong about Calvin.

[41] Torrance, 1969, p. xiii; see also Lee, 1969, p. 42.

> Hyperius) and following them by men devoted to the interpretation of the books of nature (e.g. Francis Bacon).[42]

It is Calvin, who in his *Institutes,* breaks through scholastic routine, and reverses the order of *quid sic, an sit,* and *quale sic,* so that the primary question of theology becomes *Qualis est?*[43]

> What is God? Men who pose this question are merely toying with idle speculations. It is far better for us to inquire: 'What is his nature?' and to know what is consistent with his nature.[44]

Torrance also outlines other characteristics of Reformation thought which aided the rise of modern science: the breakdown of theological support for the outdated Aristotelian scientific paradigm; the exception of the natural order from the search for 'created correspondence',[45] which freed nature to be natural and theology to be theological – a view elevated by Barth to the point that it became *the* major distinction between Roman Catholic and Protestant theology.[46] Furthermore there were of course sociological and pedagogical changes which made such developments possible.

What is also striking is that, for Torrance, those points at which the classical mechanical view is most dependent upon its inheritance from Greek philosophy are those points at which it was most changed by the new physics developed at the beginning of the twentieth century.[47]

It is often Newton (who threw in his lot with the mistaken container view of the Arians[48]), Kant, Descartes ('villains') and their followers who knowingly or unknowingly smuggled damaging dualisms into the scientific paradigm.[49] Torrance is also willing to admit that as theology sought to ingratiate itself with and conform itself to such seemingly 'successful' science, the 'hardness' of those dualisms increased after the Reformation, as neo-Protestantism settled into ever more deistic[50]

---

[42] Torrance, 1988, p. 46.
[43] See Torrance, 1988a, p. 164.
[44] Calvin, 1960, p. 41 – this is my choice of quote, rather than Torrance's.
[45] See McGrath, 2003, Vol. 3, p. 110.
[46] See Brunner in Brunner & Barth, 1946, p. 53.
[47] Support for this view: Tambiah, 1990, p. 15.
[48] Torrance, 1988, p. 19.
[49] See Achtemeier in Colyer, 2001, for the dissonance between Newton and Nicea, with the latter therefore being vindicated by Einstein. There are hints of this even in Feyerabend, e.g. KT, p. 46.
[50] This view finds support in Popper, 1992, p. 430; also Pannenberg, 2006, p. 107.

and scholastic forms.[51] These 'errors' are only put right with the advent of modern science, and it is in the field theory of James Clerk Maxwell[52] that Torrance detects the recovery of a unitary framework for natural science which will allow it to transcend its past. Torrance therefore makes much of Maxwell's (Scottish Calvinist) Christian faith. Consequently therefore, modern physics also allows and inspires Protestant theology to reclaim those fruits of the Reformation, which it had jettisoned in order to keep in with natural science.

Fluid axioms, once we have got over their oxymoronic label, do not reduce and evaporate into the open sea of Feyerabend's epistemic anarchy, because they are founded upon faithfulness to object, the experience of which cannot be dissembled away, but remains as an object in our path, to be negotiated and known. Whilst their scientific function must be always examined with an historical hermeneutic of suspicion (i.e. they are *fluid* axioms), such an examination facilitates increasingly open and heuristic encounters between traditions, so that the possibility of working effectively with fluid axioms entails that science can be weaned off Feyerabend's anarchic medicine, and the drastic epistemic side-effects of his prescription are thus exposed and avoided. Torrance is right that something can function as an axiom in a certain context, at a certain time, and in relation to subject matter, provided that we understand that in other contexts and later on, that decision may be open to critical examination, especially in the light of fresh discovery – that is scientific fallibilism – but it does not invalidate our present choice of axioms, which is justified by its fecundity. Indeed according to Polanyi – who draws attention to the local nature of the 'focal' question in hand, in which subsidiary material is necessarily taken for granted – this is how science has always functioned and can only function. A vow of objectivity (Torrance) is one of scientific faithfulness, and whilst its seriousness and problematic nature should be appreciated, it must still be undertaken, rather than dismissed as hopeless or malicious (Feyerabend).

---

51 Torrance, 2001, p. 19.

52 As Richard Kirby notes: "A shift … which owes much to James Clerk Maxwell particularly his conception of the dynamical field as an independent reality within which matter, force and field are indivisibly integrated." (Kirby, 1992, p. 9).

Evident to the senses – the wrong foundations

Feyerabend's attack on positivism and the purity of observational data undermines enlightenment attempts to found knowledge on self-evidency and sense data – itself an attempt to secure for the foundations of knowledge a high degree of inter-subjectival consensus, an undoubtedly useful tactic, and not an evangelically irrelevant one: in Acts 17:22 Paul rallies a consensus amongst his audience in order to move on to less immediately likely propositions. However as we have already seen, as we transform sense data into linguistic descriptions and move up the inter-subjective scale towards public communicability so those descriptions become increasingly framed by established theory. Imbedded within our cognitive and linguistic development is a series of negative and positive feedback loops which harden not only theoretical and conceptual frameworks but even neural pathways themselves.[53] Sense data is therefore progressively stripped of its power to be basic in the sense of noetically formational, and more likely to be variously interpreted in order to accommodate other principles – such as the constancy of the speed of light, or the first law of thermodynamics – highly theoretical principles which come to function as basic statements, *replacing* our more primitive enlightenment sense-data in the search for scientific consensus.[54]

From this launching pad however, Feyerabend derives the progression of ideas outlined in chapter five, with undue haste and sense of inevitability. For before we flee too far from the rather too simplistic schema of Enlightenment thought, and rush into the arms of an anarchic and somewhat ill-defined relativism and voluntarism, we must ask whether Torrance offers a less drastic strategy.

As we leave the safe haven of the old foundationalisms there are two obvious harbours for which we might head – other foundations, or a modified and progressive foundationalism, perhaps employing a different metaphor altogether – either of these, if successful, can rescue us from the turbulent and forbidding waters of epistemological anarchy.

[53] See for instance L. Winkley, 'Neural Pathways and the Development of the Brain' in *Primary Practice, the Journal of the National Primary Trust*, 1999/3.

[54] Scientists, given the choice between an anomalous observation or sense datum which contradicts the first law of thermodynamics, will more readily assume that the former is mistaken, and assume that the error lies there.

It is a combination of both these strategies which Torrance and indeed Barth use.

Firstly, Barth and Torrance do not simply cry 'down with the foundations' as Feyerabend does, but rather 'wrong foundations!' In this sense, Barth and Torrance simply offer a different form of foundationalism, but are forced to justify it by employing non-foundational strategies, because the theistic foundations are mediate and highly contentious – a foundational metaphysic made transcendentally manifest in a non-foundational epistemology. The New Testament itself is an obvious source of material for the Christian foundationalist, in passages such as 1 Cor 3:11, Eph 2:20, Acts 4:11, reinforced in hymnody: 'Christ is made the sure foundation', 'Christ is our cornerstone', etc.. Barth and Torrance both posit not only that, for the Christian, the revelation of God in Jesus Christ is foundational, but also that the edifice of human knowledge can *only* be built upon the foundation of that revelation. That rather bold statement demands in its turn a four fold clarification.

Firstly, they are working with a particular doctrine of revelation, to be evaluated in a moment (p. 236), which locates the foundations theistically and Christologically. However the foundations as such are *not* in a simple sense scriptural – no case is made for Biblical inerrancy and fundamentalism,[55] in fact there is outright hostility to such a bibliolatrous notion of Biblical sufficiency.[56] Torrance's Calvinism restores the critical freedom which Calvin enjoyed, over and against the increasing Biblicism of the 'Protestant Orthodoxy' which followed him.[57] Jeanrond is furthermore clear that Protestant hostility to advances in natural sciences was dependent upon the development of this 'orthodoxy', the legacy of which is seen to this day. The Biblical narrative as a whole is a necessary condition for knowledge of God,[58] but, certainly as individual texts, it is not a sufficient one.[59] Its penultimate witness must be distinguished, at the risk of bibliolatry from the ultimate to which it witnesses: "we distinguish the Bible as such from revelation. A witness is not absolutely identical with that to which it witnesses."[60]

55 Watson in J. Webster (ed.), 2000, p. 59.
56 Torrance, 1982, p. 95.
57 Jeanrond, 1994, p. 34.
58 Torrance, 1995, p. 426; Barth, CD II/2, p. 35.
59 Torrance, 1995, p. 273.
60 Barth, CD I/2, p. 463.

> Once again, the Bible is not in itself and as such God's past revelation. As it is God's word it bears witness to God's past revelation, and it is God's past revelation in the form of attestation.[61]

Secondly, there is no question of theology reassuming some sort of position as queen of the sciences, imposing its results as boundary conditions for other sciences. A doctrine of creation when stripped of any fundamentalist tendencies, offers freedom to natural science to understand the *logoi* of its own objects, and whilst theology offers a meta-account of the order and intelligibility of that creation ('coordination' is here Torrance's rather loose term[62]), Barth's insistence on the transcendence of God, grants creation its freedom from any latent divination and search for final causes. Rather natural science will function as an *ancilla theologiae* to a scientific theology.[63] Scientific theology is free to be the science of God and natural science is free to be the science of nature.

Thirdly, the diversity of objects across all the sciences entails that the various sciences interrelate in due proportion to *a posteriori* discovery of their interpenetration of the objects of others sciences, so that no science seeks to set out *a priori* the extent of its influence over foreign domains. It discovers the relations of its object to other objects, chiefly by examining its own object. The foundational role of Christology can only unreservedly play out in a scientific way with regard to the object[64] of its own science – the science of God. Thus we must not assert that Christology is foundational for the natural sciences in anything more than a strictly formal sense – the laws of motion or thermodynamics are not implicitly contained within the New Testament.

Fourthly, whilst it is beyond the scope of this book to give a general account of the Christian doctrine of revelation in all its variants, it is necessary and pertinent to point out that Barth and Torrance are working with a *particular* doctrine of revelation, and describe its main contours.

To begin with, for Barth and Torrance revelation is an event in which the prerogative of initiative, determining content and form, lies solely

[61] Barth, CD I/1, p. 111.
[62] Luoma, 2002, p. 146.
[63] McGrath, 2002, p. 122.
[64] The word 'object' should entail nothing more at this point than saying that theological science is concerned with knowledge of God, and not some ancillary or secondary concern. The objectivity of God is only a kenotic manifestation of his utter subjectivity.

with God. Revelation is therefore definitively for Torrance the event of the life, passion, death and resurrection of Jesus Christ, an event which is uniquely witnessed to by the Scriptures.[65]

> All authentic knowledge of God is derived and understood in accordance with the incarnate reality of God's self-revelation in Jesus Christ, and is formulated in doctrinal coherence with Christology. … We cannot think or speak of God truly apart from his Word and his Act in the incarnation, and that means, apart from Jesus Christ.[66]

Thus whilst, as we have seen, Torrance and Barth are not concerned with questions of the inerrancy of Scripture[67] (questions of sufficiency), on the side of necessity, it is the case that: "We must submit to our bondage to Scripture[68] … to be faithful to Holy Scripture as the only valid testimony to revelation."[69]

So Scripture is our only witness – a yardstick of reliability if not an inerrant one – to that self-revelation and its theological context, albeit by its nature mediate, and mindful also of the mutually constitutive inter-relation between Scripture and its ecclesial community. We have no *im*mediate access to revelation, therefore whilst Bultmann advocates submission to such bondage, he underlines that it must be held relatively and in tension with a critical bearing.[70] Thus analogously to natural science, the Bible is a short way along the continuum of observability, forming the most reliable data to a phenomena to which we have no immediate access. However that should not lead us to any too sharp or simplistic demarcation criteria, for the data of theological science sits on that continuum, not isolated, but in close proximity to other sciences – the human and historical being the most immediate neighbours. They too have no unmediated access to the events which they attempt to describe.

Furthermore Torrance must therefore acknowledge that there are other possible witnesses to and sources of knowledge regarding the revelation of God: the revelation of God to Israel, in particular in the Word of God addressed to the prophets, witnessed to in the Hebrew Scriptures, and the continued presence of Christ in his Church, whether

---

65 See Barth, CD I/1, p. 111; CD I/2, p. 463.
66 Torrance, 1996, p. 17.
67 Watson in Webster (ed.), 2000, p. 59.
68 Barth, CD I/2, p. 208.
69 Barth, CD 1/2, p. 280.
70 Bultmann, 1984, p. 61.

the emphasis is on charismatic, sacramental or ecclesial presence. Yet whereas Scripture, for Torrance, is something *primarily* we need to *submit* to – albeit not uncritically – those extra-Biblical witnesses are *primarily* subject to a hermeneutic of *suspicion*, as possible rivals, to an extent that it appears that something more akin to denominational prejudice rather than scientific rigour is on display.[71] Such prejudice, rather than any consideration of object, seems to determine the balance between Torrance's vows of obedience and rigour here. God's self-revelation witnessed to by Scripture becomes the ordering principle, the lens,[72] through which those other things are viewed, presupposing a negative answer to the question of whether they might be better or equally authoritative sources of Christian knowledge. So the unity of the Biblical witness to the revelation of God in Jesus Christ, becomes an interpretive principle of the Hebrew Scriptures, rather than something we discover within them. This is therefore at the cost of allowing those Scriptures to speak on their own terms.[73] Similarly faithfulness to the Scriptures becomes the plumb-line against which the historical church is measured. As McDonald says,[74] Barth finds a whole 'strange new world within the Bible', which amounted to a form of 'conceptual foundationalism',[75] so that everything is known from this vantage point.

So as pure immediate event the revelation of God is *de jure* epistemically primary,[76] however due to its mediacy, at the *de facto* level of scientific theory we take a step towards Feyerabend's relativism, for something of us is always, whether we be evangelists, bishops or scholars, going to interpose itself in that process of mediation. As Pannenberg points out,[77] whilst Torrance and Barth might order their world around the self-revelation of God in Jesus Christ, what is to stop others making similar claims for different events, lives and histories?

> Since the contents of the Christian revelation have become matters of dispute, however, it has been increasingly clear that Christian theology

[71] Torrance's 'confessional patriotism'. Louma, 2002, p. 160.
[72] Torrance, 1949, p. 41, 'spectacles' term used by Calvin.
[73] Colyer, 2001a, p. 61ff – though for balance we note that Torrance also criticises those who attempt to understand the reconciling work of Jesus Christ, without fully immersing that work in its Israelite history and context.
[74] McDonald, 2000, p. 265.
[75] McDonald, 2000, p. 115.
[76] Alston, 1991, p. 106.
[77] Pannenberg, 1976, p. 29.

> has two possible courses. It may hold on to its positivist position with an appeal to divine revelation. If it does, however, it will lose any hope of intellectual legitimisation for its claim to general validity, since it will be assuming this claim rather than proving it in a situation in which it has been challenged by the claims of other religions and beliefs. Alternatively the theology practised within the Christian tradition may see its role as Christian theology to make the superiority of Christianity to other systems of belief the explicit object of investigation and proof in a theology of religions.[78]

Torrance is consistent, in that he admits this 'relativity'[79] at the epistemic level, but denies the possibility of the philosophic-ontological relativism inherent in much post modern thought,[80] which seeks to drive epistemic relativity down to the ontic level ('mirrors all the way down'[81]). As we have noted this is founded on the a-theistic assumption that once our description of the world is shown to be relative and particular, there is no other being who might grasp ontology 'in-itself' – the God-view simply does not exist.

Torrance, following Barth, also denies the possibility of assessing the revelation on any neutral criteria,[82] as though we might sit in judgement over God's revelation.

> From 1915 onwards Barth's objective on epistemology remained unchanged: to formulate exactly what it meant to be in a position where one could not take up a position on God independently of God's self-revelation itself.[83]

Neither is there therefore any possibility of *proving* the existence of God[84] in a traditional sense. "Note well: in the whole Bible of the Old and New Testaments not the slightest attempt is ever made to *prove* God."[85]

What Torrance comes closest to is Moore's transcendental argument for (specifically) Christian realism, whereby what is outlined are the

[78] Pannenberg, 1976, p. 322.
[79] As distinct from relativism, this term is here used as it is in general relativity theory.
[80] Schwobel in Webster (ed.), 2000, p. 33.
[81] Rorty, 1991, p. 100.
[82] Barth, CD II/1, p. 5.
[83] McDonald, 2000, p. 67.
[84] Barth, CD I/1, p. 30.
[85] Barth, 1949, p. 36–7; see also CD I/1, p. 30; CD II/1, p. 32 – in the Scottish tradition this point is found more or less word for word in Torrance's teacher: Norman Kemp Smith in Fergusson (ed.), 2007, p. 172.

conditions upon which such realism depends.[86] The 'prevenience' of God's reality and revelation in Jesus Christ is thus preserved, which is why, as Putnam observes,[87] medieval philosophy had far fewer questions over realism than its contemporary counterpart.

Torrance's justification for this Christocentric particularism is personal conviction and coherence.

> Through my missionary parents I was imbued from my earliest days with a vivid belief in God. Belief in God was so natural that I could no more doubt the existence of God than the existence of my parents and of the world around me. I cannot remember ever having had any doubts about God. Moreover as long as I can recall my religious outlook was essentially Biblical and evangelical, and indeed evangelistic ... this orientation to mission was built into the fabric of my mind, and has never faded.[88]

Luoma, not unreasonably, suggests that such reasons cannot be decisive in scholarly debate.[89]

Torrance's particular theological schema has attracted considerable fire from his critics, whose criticism include: the possibility of a more positive assessment of natural theology; his too easy distillations of a single Biblical theology, and of a single classical tradition within Christian theology[90]; a slight pneumatology[91]; an unduly negative assessment of the contribution of ecclesial and magisterial theology – Torrance is keen to point out that the church does not partake in the divine nature in the same way as Christ,[92] and thus tends to downplay the Biblical and apostolic continuities between Christ and his church; working with a particular doctrine of Scripture outlined above.[93] A comprehensive survey of his theology is not being attempted here, so we have to content ourselves with considering these issues only in passing. However the main point to be made with regard to the

[86] Moore, 2003, p. 130 "It attempts to bring out the conditions and assumptions which makes this realism possible and in this way amounts to a transcendental argument for a Christocentric realism."

[87] Putnam, 1987, p. 52.

[88] Quoted by McGrath, 1999, p. 13, from '*Itinerarium mentis in Deum*'; Torrance's colleague at Edinburgh, John Baillie, recalls a similar starting point, in Fergusson, (ed.), 2007, p. 184.

[89] Luoma, 2002, p. 123.

[90] Hardy in Ford, 2005, p. 173, also Norris, 1984, p. 466.

[91] For instance Robert Jenson, quoted in Spjuth, 1995; also Nuoma, 2002, p. 128.

[92] Torrance, 1969, p. 213f.

[93] Barth, CD I/1, p. 274 – questions of Bible's truth "can only be put as questions of obedience."

scientific status of theology is that, given this cannot be determined apart from a theological identity, this therefore means that that status is affected by the particular sort of theological schema a particular theologian is advocating – because this question cannot be resolved *a priori*, so the question of scientific status will vary between theologians.

Logically if we assume the unity of God's purpose and being, however iterative our procedures, our starting point becomes irrelevant – convergent iterative series converge to the same limit however far along the line you start, so that the imbalances and particularities, traditioned by denominational concerns, of which Torrance is guilty, will eventually be corrected as part of that process. However given that Torrance is very clear that the intuitive 'integration of form' is a necessary part of rigorous *scientific procedure* – this intuition going beyond algorithmic and logical functions – so that procedure demands that we start with as open schema as possible, working up with fluid axioms to particular commitments carefully and contingently. As Polanyi points out[94] good initial 'hunches' are part of the economy of good scientific method – starting with wild ideas, is likely to lead only very slowly away from wild results in iterative processes. In one sense this is Torrance's point, yet his confessional outlook is so visible even within those initial 'hunches', that we cannot but help conclude that the circle has been drawn rather too tightly. Furthermore the object of theology dwells in eternity, and sees human history through eschatology, so that the scope for repetition of the iterations within theological controversies may well seem from our point of view as being infinite – a brief glance at the history of doctrine confirms this as true! No doubt in the kingdom all will know and all will agree, but meanwhile theologians are working to the timescales of our present age.

As Jeanrond therefore correctly points out, in this case with reference to Barth,[95] there is no reason "why an open discussion of the presuppositions of any approach to texts would need to be seen as *determining* our understanding of these texts." Theological and denominational commitments may determine and will surely *condition* our reading, but there is a big difference there, to be negotiated simply by an open and honest reading. All the while we need to heed Feyerabend's reminders that in Barth's time, natural science did, and sometimes

[94] Polanyi, 1967, p. 29.
[95] Jeanrond, 1994, p. 137.

continues to, use strong armed tactics when it is one of the dialogue partners, thus hampering the search for open dialogues and the 'open discussion' Jeanroad appeals for.

Therefore we should not underestimate the difficulty of the task, and Feyerabend is implicitly tempting Torrance to give up the attempt because of its scale. If we begin the dialogue at Torrance's starting point then the whole narrative of modernism, which still powerfully holds sway in many areas, will have to be re-written – 'Christianised' ("the revolutionary task of a Christian reconstruction of the foundations of culture; nothing less is worthy of the Christian gospel"[96]) – simultaneously resisting Feyerabend's temptations, indeed not acting on any a-theological philosophical prescriptions, but operating from within the 'strange new world of the Bible'. The question of whether that project is theoretically possible or not, seems dwarfed by the fact that it is unfeasibly difficult, and that Jeanrond's modest suggestion that Torrance and Barth should be willing to play along and listen a little more is an undeniably reasonable one.

What Torrance is attempting to avoid is the perennial pitfall of foundationalism – how do you defend the choice and validity of foundations unless they are self-evident? The temptation is to try, but by so doing you undermine their foundational status. Feyerabend has pointed out that the senses are unsuitable, and goes on to conclude that this problem is crippling for any reformulation of foundationalism. Torrance seeks to show that for the theist and in particular for the Christian, any foundation other than the theological and Christological is not only unreliable epistemologically but impious, for it is likely to usurp the primacy and sovereignty of God.

Barth and Torrance's reassertion of the pre-eminence of revelation over reason reverses the Kantian revolution.[97] "The crux of the matter for Barth is Kant's affirmation of the logical prevenience of reason over revelation."[98] Though generally such generalisations stand or fall on precisely what one means by reason and revelation, and perhaps more importantly whether it is possible to draw any sharp distinction between the two. Torrance invokes (in a slightly different context) the

[96] Torrance, 1965, p. 271.

[97] Kant, *What does it mean to orient oneself in thinking*, 1998, p. 9–10; Kant, 1998, p. 100; Kant, 1979, p. 71.

[98] McDonald, 2000, p. 116; see also Moore, 2003, p. 123.

Aristotelian lesbian rule[99] to demand that reason should reform itself in the face of the facts of revelation. Indeed rationality demands of itself, that its own relation to the facts as known, be rigorously investigated, and that a crude assertion of reason's primacy ironically offends against its own nature – it is "an act of sheer irrationality."[100] As Moore says, (Torrance's) Christian realism is "fideism of a second order: reason-giving, rather than a reason-less exercise."[101] The lack of warrantable logical connections between epistemic and metaphysical levels, taken by post-modernism to imply their absence,[102] is interpreted by Barth and Torrance simply to indicate that here we must live by faith – second order fideism of the reason-giving kind, not capricious first order fideism. The Gospel is extra-logical not illogical[103] – its connections go beyond the proper scope of 'paper logic', and will not submit to Feyerabend's demand that in order to be rational "they must have an [logical] argument for every move they make." Ultimately Feyerabend's straw man rationalist is not very rational at all, being too stiff and lifeless – that is both Feyerabend's point and his weakness – once we have breathed some life into him, then he is immune from Feyerabendianism. However the 'charge' of fideism is therefore one that is always likely to be levelled at Barthian theology:

> Either Barth's epistemology is apologetic or it is fideism. Since it is not the former, it must be the latter. Does this mean that the Enlightenment, on the side of apologetic, can dismiss Barth to the realm of fideism as it (counterwise pejoratively) understands this position. No. … Barth had another epistemological option to offer, an argument in the old fashioned sense, in the form of conceptual foundationalism.[104]

The truth of the Biblical material is to be known by its correlation as witness to the *veritas Dei*, the revealed Word of God – the Ignatian notion of the Word as that 'in which God broke his silence', as opposed to, on one hand the 'time to time' prophetic word, and on the other the mainstream patristic *logos* theology.[105] There can be no other ground on which we justify the reliability of the witness of Scripture, other

[99] Torrance, 1988, p. 117.
[100] Torrance, 1969, p. ix.
[101] Moore, 2003, p. 135.
[102] Colyer, 2001a, p. 324.
[103] Colyer, 2001a, p. 337.
[104] McDonald, 2000, p. 305–6.
[105] Pannenberg, 1968, p. 167.

than Scripture itself, and though the church creates the canon,[106] the church henceforward is judged according to the standards of the Biblical witness – in his political situation, it was vitally important to Barth that the church was known to be the church under the judgement of the Word of God,[107] in its three fold witness of revelation, scripture and preaching. So for Barth and Torrance the Bible's reliability as witness can only be known from within the Bible as a whole, it cannot ultimately be judged on independent historical critical grounds. Neither can we seek to construct a 'truer' historical Jesus, or a demythologised gospel, behind the actual Biblical material.[108]

According to Torrance the interpretation of the Bible, though it can be informed by and must engage with our fore-conceptions, cannot be forced to conform to those conceptions, but must be self-interpretingly allowed to speak to us on its own terms, whatever our pre-judices and pre-conceptions may be.[109] In so far as we do accommodate, perhaps unconsciously, Biblical interpretation within an extra-Biblical framework, a premature failure to take seriously the Biblical claims over against our own horizon will be distortive of any act of interpretation. This was the error not only of the (old) historical critical school, but also the Gnostics,[110] Thomists, and demythologizers such as Bultmann.[111] This marks both Barth (paradigmatically) and Torrance out as 'type four' theologians in Frei's typology.[112]

Barth is fully aware that thereby we set the task of theological science within a circle, but according to him not a logically vicious circle, rather the circle of God's truth, from which in any case there is no escape, for its orbit is by definition all encompassing:

> this circular course of our presentation is not in itself a proof that we have really established and understood the truth of the knowledge of God that takes place in faith …

[106] Barth, CD I/2, p. 601, is prepared to admit that the canon is still 'open' in a sense, analogous (my analogy) to a well-established unfalsified theory in science, which is however still, theoretically at least, vulnerable to refutation.

[107] Barmen declaration, 1934, item 1.

[108] MacDonald, 2000, p. 224.

[109] As per Gadamer, 1975, p. 236.

[110] Torrance, 1995, p. 106–7.

[111] Torrance, 1971, p. 79.

[112] Frei, 1992, p. 4; McGrath detects the roots of this approach in "Luther [who] opposes those who smuggle an *a priori* metaphysical system into theology, not those who derive such a system *a posteriori* on the basis of an engagement with Scripture." (McGrath, 2003, Vol. 3, p. 278).

admitting that:

> the determination of the limits, the question of the veracity of the knowledge of God, presupposes the very thing which we there tried to elucidate: that knowledge of God is real and possible. '*Circulus veritatis Dei*' rather than '*circulus vitiosus*',[113]

in a direct echo of Heidegger,[114] and in the Scottish tradition, of Edward Caird.[115]

In CD II/2 Barth answers the fundamental question "why is God's revelation true" with the reply "because it is God's":[116]

> He has no means of measurement of *what it is* that God's self-revelation is, other than God's self revelation itself. The fact that there is no such means of measurement other than the object of measurement itself is testimony to the Godness of God. Man must set himself aside and rely utterly on God revealing himself: 'He must increase, while I must decrease'.[117]

This move is judged simply illegal by Geldhof,[118] though his reasoning is only valid within the deductive framework of ground and content. Furthermore[119] this construction of a revelatory 'untakeable fortress' (a veritable *Feste Berg*), equates simply, according to Jaspers, to "untruth" – that is simply false, for even within the ground / content scheme, true conclusions can follow invalid arguments.

Torrance also takes this circular path,[120] (which we have already argued in our discussion of hermeneutics is common across all sciences) likewise rejecting the implication that it is vicious. Theology, he says, at times "involves us in a proper circularity."[121]

It is not hard to see that whereas we started by asserting the foundational credentials of Torrance's theology (ontically), we have had to move to a defence of a 'proper circularity' (epistemically).

---

[113] Barth, CD II/2, p. 242, 246. Leading Brown to conclude that: "never was dogmatics more self-consciously set within the 'theological circle' of which Paul Tillich speaks." Brown, 1955, p. 143; see also Moore, 2003, p. 10.
[114] Heidegger, 1996, p. 152.
[115] Caird in Fergusson (ed.), 2007, p. 141.
[116] Barth, CD II/,1 p. 209; CD II/2, p. 242, 246.
[117] McDonald, 2000, p. 347.
[118] Geldhof, 2007, p. 10.
[119] Geldhof, 2007, p. 28.
[120] Torrance, 1976, p. 14.
[121] Torrance, 1969, p. 36.

The 'mediateness' of the Christian revelation for Torrance[122] and Barth,[123] means that the foundations are not something which human science can 'get at' and that those foundations are therefore contentious – historical distance, the personal nature and mysterious sovereignty of God, the 'blindness' and 'deafness' (e.g. John 12:40) of many of the witnesses of revelation, all combine to make this problem sharper than when we are concerned with knowledge of the *Ding an sich* of natural science. Thus we must admit that Pannenberg and Torrance offer us a genuine dilemma – the hope of a degree of consensus, versus theological fidelity.

Torrance's approach is unhelpfully narrow here, for it introduces a particularly sharp iteration in the spiral of knowledge, in that the foundations, to put it bluntly, appear to have little more rational force than an accident of birth or at least upbringing.[124] This is a different sort of openness than the parallel one Torrance is describing in natural science, though it shares some of the same features. It is one thing for the particle physicist to hold subsidiary as a fluid axiom the 'atomic' hypothesis, in his investigation of quarks, and quite another for the theologian to hold a high doctrine of incarnation, imbued as part of his inherited faith, as part of the framework within which he engages on pressing matters of inter-faith dialogue – physicists simply do not 'imbue' the atomic theory unquestioningly from their parents in the same way. It would seem valuable, if only as a 'thought experiment' – a natural scientific technique unhelpfully absent from Torrance – for scientific theology to consider, more than Torrance allows, what it would materially entail if, at various points, it allowed itself to be examined by those criteria which would offer it more general validity. Even if it doesn't therefore wholly commit to Pannenberg's search for general validity, can it not sometimes play the Pannenberg game, like a scientist tries out a model or idealisation to see where it gets him?[125]

---

[122] Torrance, 1995, p. 426; 2001, p. 6; 1969, p. 106.

[123] Barth, CD II/1, p. 16; CD I/2, p. 540; see also Torrance, 1990, p. 163; Brown, 1955, p. 152.

[124] See quotation from '*Itinerarium mentis in Deum*' in McGrath, 1999, p. 13.

[125] It should be noted that Bultmann too was reacting against the strictures of the historical critical school, and his appeal that we get our hermeneutical presuppositions out in the open is a mark of good method, and not necessarily binding theology to prior commitments it cannot fulfil. If that is what demythologising became, or if Bultmann was too much tied to Heidegger's philosophy, terminating his inquiry with questions of human existence, then that, as Jeanrond points out, is him simply not seeing through his hermeneutical proposals to their conclusion, wherein God is known

The danger for Barth and Torrance is that the nature and integrity of theology would somehow be of necessity compromised or bound, in a way which thought experiments should not do. However that is precisely the all-or-nothing approach already highlighted in Feyerabend – part of scientific fallibilism is the willingness to play *other people's games*, even if scientific rigour also reserves us the right to cross our fingers behind our back. Part of the responsibility for keeping dialogue open lies with both of the participants. There is a genuine danger in the 'anthropological method' whereby one is fooled into thinking that one has stopped being an anthropologist and has actually 'become' an Azande – but that danger does not necessarily mean that anthropologists should not seek to immerse themselves as fully as possible in the cultures which they are studying.

The danger of Torrance's reconstructed foundationalism is that the foundations are hidden under such calcified human mediation that the visible edifice is very like the capricious construction of the relativist, and as we have seen, in some places Torrance is forced to defend it as such. Within that defence, the dignity of God's sovereignty and mystery is safeguarded by Torrance only at the cost that God becomes a hypothesized meta-theoretical being, the ultimate of Torrance's 'elemental forms'. If his existence will not condescend to human interrogation, then we are likely to be at the mercy of Feyerabend's theory-determinedness, for is theism not the theoretical commitment whose self-referential mythic primacy trumps all others? As said earlier we detect that Torrance is anxious to protect God's sovereign subjectivity in a semi-Pelagian sort of way, as though God's being and existence should not be risked in scientific investigation, lest some offence to his dignity is thus caused. At the risk of impiety one might suggest that God's subjectivity is not so fragile that it needs such protection – if we errantly commit ourselves to inappropriate conceptualities that is no doubt another sign of our frailty with which God will have to learn to live, and forgive.

* * *

So whilst the senses might not play the foundational role which the Enlightenment would have preferred, Torrance is keen nonetheless to

---

"as something which lies in my way … it is not protected against an easy dissolution into my mere opinion of what it represents." Jeanrond, 1994, p. 146.

underline the positive role of physically embodied data[126] – of an historical form – within Christian knowledge.[127] For here the Christian is at an advantage over other forms of religious belief, because (and we are assuming that by 'Christianity' we mean something like the Torrance form) knowledge of God is made known by revelation which has certain physically embodied manifests in the life, death and resurrection of Jesus Christ. Whilst this is far from saying that Christianity is reductively 'evident to the senses' or 'verified by observation', it is equally distinct from saying that it is purely based on some form of extra-sensory data. If history is a science, then the history of Jesus Christ, and the documents of the New Testament admit scientific investigation.

Yet Barth and Torrance are ambiguous about the efficacy of the historical critical method within Biblical studies, particularly as Protestant theology emerged from the nineteenth century somewhat 'ensnared' within the commitments of a positivistic style of criticism.

Let us start by noting that Barth flatly denies that sense data can provide an adequate starting point for the doctrine of God.[128] This we must affirm. Jesus' detractors and enemies were not so because they did not have a clear view, or because they could not hear him clearly (Luke 19:47).[129] Conversely there remains an existential component to the decision demanded by the gospel today, just as amongst Jesus' contemporaries.[130] It is existential rather than historical distance which principally separates followers and despisers of Christ. The gospels are free from what we might call more 'objective' sense data about Jesus Christ – for instance there is no description of what he looked like. The gospels are theological witness, not neutral ones certainly, but generally 'impartial' witnesses (as opposed for instance to later Gnostic accounts) which seek to capture the reality of the object, the revelation of God in Jesus Christ.

So theologically coloured an object demands in and of itself theological rather than simply historical description, anything other would not be true to reality. The points of substance with which the gospel are

[126] Colyer, 2001a, p. 189.

[127] As opposed to Bultmann, who downplays them "God's eschatological act is not visible to existence outside of faith," 1984, p. 66.

[128] Barth, CD II/2, p. 3.

[129] Holwerda in Plantinga & Wolterstorff, 1983, p. 286 – criticism is of Pannenberg.

[130] Kierkegaard, 1946, p. 131 "There is no disciple at second hand"; see also Frei, 1975, p. 172.

concerned are simply not those which can be explicitly derived from observation data. As Rae points out, this marks an important advantage of Barth's approach over Pannenberg's with regard to the relation of theology to history: "there is nothing to be observed in the history of Jesus of Nazareth that confirms him as the Son of God,"[131] – indeed it is difficult to imagine what such confirmations might consist of – therefore "I agree, contra Wolfhart Pannenberg, that theological claims, like the one that Jesus is the Son of God, cannot be set forth as straightforward deductions from observable events …The truth of history cannot be reduced to the raw data of perception."[132]

Whilst we will come to evaluate in more detail particular elements of historical criticism, Torrance's point at this stage is that if the aim of such criticism is to allow the reader to assume the place of Jesus' historical contemporaries, then we already know that such an aim is religiously inadequate – most of Jesus' contemporaries, including good religious people, regarded him as an imposter and a disgraced and probably deranged apocalyptic preacher – only by allowing the events of revelation to elicit from us a response of repentance and renewal can we seek to know Jesus as the Christ of God.[133] Only within the decontaminated context of the proto-community constituted by the Messianic secret can that secret be made known. Pannenberg criticizes 'christologies from above' which 'presuppose' the divinity of Christ,[134] but this implies a linear hermeneutical structure moving from data to theory (not unlike Feyerabend), in which presuppositions 'determine' results, whereas the decisive point is not *when* they are introduced into the discussion (i.e. from above or below), but *how* rigorously such 'intuitions' are examined and justified. If Christ's divinity functions unquestioningly either as a prior commitment or a post-hoc justification, scientific rigour is undermined; it can only be known within the spiral of our encounters with the text.

The data of the Biblical material is, as Feyerabend would say, theory laden, not only in the particular theological beliefs and world view of its writers, but in the creation of the canon by the church, and its subsequent interpretation in differing ecclesiological settings. These processes of creation and transmission are by no means only

131 Rae, 2005, p. 28.
132 Rae, 2005, p. 87-8; see also Farrer, 1948, p. 43.
133 See also e.g. McCabe, 2007, p. 153.
134 Pannenberg, 1968, p. 34.

or even primarily ideational, but as Feyerabend would point out, questions of power, politics, prestige and wealth play a significant part. Many would therefore take the line that the Biblical witness is so strongly theory-infected that it is unreliable to the point of uselessness – but Torrance would insist that such 'theory-ladenness' is on balance faithful to the reality it seeks to depict – there is also, slightly more negatively, no viable alternative. Certainly the relation of data and theory, of Biblical witness and systematic theology, cannot be hermetically sealed into a logical positivist division of observation and theory – a tendency which theology in the nineteenth century was particularly guilty of.[135] Neither can we rectify this with new experimental results. The Biblical witness is, according to Torrance, *properly* theological – faithfully theory laden, rather than mythically theory determined – given that its object is the revelation of God, how could it not be the former if it seeks accuracy of depiction?[136] Systematic theology can clarify the proper interpretation of Biblical texts, yet cannot itself be constructed apart from reference to that text as a whole. There is therefore not a logical one way street between data and theory, but an iterative and spiralling interaction between the two. However we should not underestimate the scale of the task Torrance sets us in order successfully to negotiate the unscientific intrusions within the Biblical theology as a whole – encompassing at least the disciplines Biblical studies (including the history of interpretation), historical theology, systematic theology, and church history – a task which Torrance can only point the way towards.

Whereas Feyerabend warns us that data over-laden with theory determines the data, Torrance warns us of the obverse problem, namely that data yields nothing meaningful without proper interpretation. The temptation within historical science to mimic natural sciences and attempt to screen off fact from value or interpretation was not only impossible, but also undesirable even as a limit point, and likely severely to restrict the fruitfulness of such historical investigations – the recognition that such an attempt was undesirable was the distinguishing feature of the so-called 'new' search for the 'historical Jesus' – Barth and Torrance were reacting against the crude positivism of the old.

On one side of the coin, these historical correlates can only be understood using the tools of historical research, so that the use of historical

[135] Morgan, 1988, p. 175.

[136] Thus as we shall see the construction of a non-theological 'Jesus of history' is ultimately an unscientific infidelity to object.

methods is not only licensed but demanded by the text itself. On the other, the incarnation understood within the context of salvation history marks a singularity which cannot be understood solely within the conceptualities and categories of secular historical research, thereby questioning those conceptualities to their very core. We have no option but to use theological concepts and categories properly here, for the text itself demands it. This is not (in fact it is the opposite) of dividing the Jesus of history from the Christ of faith – for only within its own unique orbit as revelatory event can the unit Jesus-Christ be understood – his history as human person can only be understood by relation to his divine nature.[137]

In terms of genre, the gospels are therefore difficult to assess because of the disjunctive temptation either to place them simply in a class of their own – thereby cutting them off from the tools and resources used in the study of other texts – or to force them into a Procrustean bed inappropriate for good understanding.[138] They are, to use Frei's phrase, '*historylike*'[139] documents – Burridge compares them, against the trend, with Graeco-Roman biography.[140] They invite historical research, but only if the researcher is able and willing to enter into a dialogue in which those methods are questioned, just as the good workmen does not blame his tools, but continually questions whether they are the appropriate ones. Particularly when theology is concerned about its place in the secular university, there has been a tradition (Meier – Schleiermacher – Pannenberg) of treating the Biblical narrative in the same way as any other text of antiquity.[141] This is the point at which Barth parts company from the liberal-Protestantism of his teachers, Troeltsch and von Harnack.[142] Barth believed that within the methods of historical science[143] certain unchallenged presuppositions (see following) would threaten theology's autonomy and integrity (similar to

[137] John Baillie (Fergusson (ed.), 2007, p. 187) states that this dichotomy is in fact Barth's – that is not my reading of Torrance or Barth.

[138] Frei, 1974, p. 274, "They had no category for dealing with the Biblical narratives other than the disjunctive device that assigned either ostensive or mythical status to them."

[139] Frei, 1974, p. 10ff.

[140] Burridge, 1992.

[141] Gadamer, 1975, p. 290, with particular regard to Schleiermacher at this point.

[142] See Schwöbel in Webster (ed.), 2000, p. 23.

[143] This is of course an ever evolving discipline, but the end of the nineteenth century marked the high water mark at which the humanities and human sciences, dazzled by the successes of the physical sciences, tended to view the methods of these

the manner in which the findings of natural science as set forth by Scholz were liable to 'ensnare' theology), and that therefore whilst such methods might be useful, they tended both to be distortive of the meaning of the Biblical text, and apt to miss its true significance and reference. Barth and Torrance therefore both offer the same verdict on the importance of the historical critical method. Barth:

> This task is performed by the scientific study of the Bible, which in recent times has developed into what is called the historical and critical study of the Bible. This is just the point where it is important for us to note that neither too little nor too much should be expected from such a study. One is entitled to expect from it that it will clarify the whole human form of the witness to Christ in the Old and New Testaments, throwing light on its linguistic, literary, historical and religious-historical aspects. But we should not expect it to set before us the object of this testimony, which is God's revelation and therefore Jesus Christ as the Messiah of Israel and Lord of His Church. How could revelation ever be recognized as the divine content of that testimony except through revelation.[144]

Torrance:

> Such is the Jesus of history who has haunted and baffled so many modern historians, for He insists on being understood as true historical event that will not submit to historical criteria we already have at our disposal, but will yield His secret to those who are open and ready to consider Him in His own majestic self-presentation and self-interpretation.[145]

This verdict has not been universally accepted, particularly when the concern is with the place of theology in academic life. Here the demand for 'public' non-confessional and therefore non-immanent historical-scientific standards has been upheld, notably by Pannenberg:

> A somewhat more conservative theological [Barth's] tradition holds with Pannenberg that the historical is the *conditio sine qua non* of faith but maintains that the revelatory character of these historical events cannot be known by historical reasoning. Against this position Pannenberg continues to maintain that if 'the revelatory and redemptive significance of the fate of Jesus of Nazareth can be seen only by faith and is in principle

sciences, particularly Pythagorean quantifications, as both worthy of imitation and as setting unbending limits for their own results. Both Feyerabend and Torrance believe that other traditions are liable to compromise their own proper integrity – a methodology correlate with their own objects – by this imitation.

[144] Barth, 1938, p. 66.

[145] Torrance, 1969, p. 335.

> closed to the rational investigation of this event, then it is impossible to see how the historicity of the pure facts should be able to protect faith against the reproach that it rests upon illusion and caprice.[146]

However the possibility or profitability of the recital of these 'pure facts' is precisely one of the points at issue.[147] As we stated above, the intermingling of theory and facts over which Feyerabend forces natural science to sweat, may be crippling to a positivist construal of the scientific enterprise, but is positively mandatory within the human and historical sciences. The terms, concepts, names, distinctions, and sources used within history are fully theory laden. All historical fact is theory-mediated. Thus any attempt at a false objectivity borne out of too strong a convergence between the methods of physics and history is brought up short by physicists themselves in quantum theory:

> In essence the "Heisenberg principle" or field theory in more radical form may be seen in operation here; that is, that the object being observed is itself subtly altered simply by the condition of being observed. The historian is part of the very field he is observing. Historical knowledge is itself an historical event; subject and object of historical science do not exist independently of one another.[148]

Frei traces this principle in historical hermeneutics ultimately to Calvin.[149] In fact it is more closely parallel to the measurement problem in quantum physics, and as in physics, what is more important is an assessment of the significance of the effects such problems entail. Not all physicists are frozen in a state of perpetual epistemic angst on account of the counter intuitive aspects of quantum theory, because they develop an instrumental understanding of how significant such effects are likely to be. Similarly historians and hermeneuts are able to

---

[146] Holwerda in Plantinga & Wolterstorff (eds.), 1983, p. 285; see also Pannenberg, 1968, p. 109.

[147] "That is the recurring denouement of the quest of the historical Jesus which for a hundred and fifty years has been carried on within a framework of radical dualism of fact and meaning, or empirical and theoretical components of knowledge, necessarily creating an unbridgeable gulf between the original Jesus and the theologising tradition of the Church. Moreover such a procedure inevitably predetermines the results of historico-critical research from the very outset, … into the axiomatic assumption that at all states it is man's knowing of phenomena which gives them the intelligible shape and structure in which their meaning is claimed to lie." Torrance, 1984, p. 246.

[148] Palmer, 1969, p. 52; also Bultmann, 1984, p. 136.

[149] Frei, 1974, p. 36.

develop effective strategies to minimize and allow for their influence on the texts and events which they are studying. Thus in history:

> when background knowledge and knowledge of facts fuse to such an extent that the facts are no longer able to service as judge over our interpretations. Stegmuller characterizes this fusion as the 'theory-ladenness of observations'.[150]

So Biblical studies will properly seek to quarantine the theological excesses and enthusiasms of particular Biblical writers by placing their accounts within the perspective of their own particular contexts,[151] and the results thus derived possess utility for the theological scientist, not as positive results for the science of God, but negative ones which seek to identify human occlusions and accretions within the Biblical witness.

Forms of historical criticism which submit the Biblical text to demythologizing, beg the question as to whether corresponding mythological (a favourite Feyerabendian concern[152]) or indeed simply false components lie within our own scientific horizons, which thereby force a Procrustean pruning of the Biblical narratives (such as Scholz' congruity postulate). As Torrance points out such errors then "predetermine the [errant] results of historical-critical research from the very outset."[153] This is certainly a fair complaint in practice, but that is not to say that historical search will necessarily 'predetermine..*at the outset*', as long as there is a genuine dialogue in which the text is allowed to challenge those assumptions. It is also vital that not only must theology not 'marry' the current paradigm, but also it must keep up with what the current paradigm actually *is*,[154] whilst of course admitting that the pre-modern outlook of Biblical writers may equally well be determinative for them.

For instance, classical Newtonian science intuited that space and time was homogeneous (the cosmological postulate),[155] by analogy

---

150 Connolly and Keutner in Conolly and Keutner (eds.), 1988, p. 42.

151 Watson in Webster (ed.), 2000, p. 66.

152 Feyerabend, AM, p. 298; Feyerabend, KSR, p. 60.

153 Torrance, 1984, p. 246.

154 Whether this in turn ensnares Torrance's theology within a world view as particular as that of Newton (unlike Barth, who just steers clear of such questions) is as we have seen a charge laid against him by Weightman. The point remains that between Barth and Torrance significant developments within natural science occurred.

155 Charon, 1970, p. 192.

therefore so was history.[156] However for Christian theology the incarnation is a spatially and chronologically particular event, and time and space are therefore not homogeneous. "Western thought from ancient times in Greece has developed the habit of getting rid of singularities through some kind of scientific horror of unique events."[157] Troeltsch's principle of analogy is for Barth inapplicable to the event of revelation, because the whole force of such an event is that it is unique and particular, and cannot be understood by such a principle.[158] However subsequent to Troeltsch, the analogy also breaks down from the side of natural science, because within general relativity theory, time and space are themselves no longer homogeneous. More recently we note that Cupitt joins this principle with Scholz' congruity postulate thus: "There are overwhelmingly strong reasons of a scientific kind for supposing that in first century Palestine the laws of nature were just the same as they are now."[159] He neither states those reasons, nor describes why those laws rule out theological singularities.

Spirals and iterations – the search for a new metaphor of knowledge

Therefore whilst Torrance's outlook is ontically foundational, the mediate and mysterious nature of the knowledge of the bedrock, renders it epistemically non-foundational – *our spades are turned not by bedrock, but by our reliance on God's grace.* We may never have the epistemic strength to reach the bedrock, but the evidence for its presence is given in the stability and rationality of the created order in and on which we exist. However in order to avoid Feyerabend's anarchic pull, Torrance has offered another sort of structural mechanism – fluid axiomatics – and now we must see if this structure can be given any metaphorical shape.

Popper, working within the foundational metaphor, describes a less well founded structure:

> Science does not rest upon solid bedrock. The bold structure of its theories rises, as it were above a swamp. It is like a building erected on piles.

[156] Pannenberg, 1968, p. 116; Newton–Smith, 1990, p. 229.
[157] Torrance, 2001, p. 100.
[158] Barth, CD II/1, p. 75ff; Troeltsch, *The Absoluteness of Christianity and the History of Religions*, Westminster John Knox Press, 2006, p. 9.
[159] Cupitt, 1980, p. 44.

> The piles are driven down from above into the swamp, but not down to any natural or 'given' base; and if we stop driving the piles deeper, it is not because we have reached firm ground. We simply stop when we are satisfied that the piles are firm enough to carry the structure, at least for the time being.[160]

Wittgenstein similarly suggests that we need to identify the points, not of ultimate foundation, but the penultimate points "at which my spade is turned"[161] – the distinction being that we are conscious that this may not be a permanent halt, but that which must suffice as axiomatic for *now*.[162]

It is upon this 'marshy' ground of fluid axioms where Torrance is encouraging us to build, rather than upon Feyerabend's open sea. However even here, to stretch the metaphor, human ingenuity can create water-tight structures – we are not simply left clinging to driftwood. When building on marshy ground, builders talk of laying down a 'raft' upon which the building can sit. The metaphor has an epistemological parallel in the notion of the framework of knowledge being more like a boat than a building, an idea usually associated with the work of Neurath.[163] It is also taken up as a metaphor by Bhaskar,[164] an important influence on McGrath who sees the metaphor's theological potential and resonance:

> The implications of this for a theology based upon the Christian tradition are significant. Theology takes place within a tradition which is already in existence. The journey of faith has already been begun. The Christian tradition does not require to be constituted; for that has already taken place at its formative stage; it requires to be consolidated and developed. Yet we are not forced to accept the tradition precisely as it has been given to us; we are able to make 'running repairs' as we progress, in the light of our experience of the journey thus far, and the weaknesses and difficulties which this has exposed. Neurath thus uses the image of the boat to develop a philosophical equivalent to the theological notion of 'living tradition'.[165]

Griffith-Dickson[166] traces the progress of this metaphor from Kant, through Clifford to Plantinga and Lorenzen.

---

160 Popper, 1992, p. 111.
161 Wittgenstein, 1967, p. 217.
162 Putnam, 1987, p. 85.
163 Neurath, *Empiricism and Sociology*, 1973, p. 198.
164 Bhaskar, 1991, p. 13.
165 McGrath, 2002, Vol. 2, p. 35.
166 Griffith-Dickson, 2000, p. 9ff.

People can live and find meaning and purpose within a building or a boat, as they can within a 'living tradition'. They are not simply epistemic structures – a tradition is a form of life – '*Theologia* is *Eusebeia*.'[167] Feyerabend therefore wishes to judge traditions not on purely epistemic grounds but by the criterion of human flourishing. However, particularly within his democratic decision making framework, this is liable to lead to a dangerous short-termism, just as dangerous as an imperialistic epistemology – proper regard for epistemological concerns may promote human flourishing more effectively in the long term. The danger of Torrance's scientific prevenience is conversely that our tradition may be a rather sterile and uncomfortable living space, and we would do better to tolerate a bit of dirt to strengthen our noetic immune systems – after all scientific knowledge is only part of the living space which is a tradition.

As with any metaphor, we can also ask further questions: who decided where the boat is going? If we are badly damaged can we head for shore to carry out repairs? The most crucial question we must put to Torrance and particularly Feyerabend is what happens when boats meet – are these friendly cruise liners, where people can embark, and disembark, swap boats, cruise together and communicate, or are they Feyerabendian gunships patrolling their territorial waters, and assuming battle formations whenever they encounter another vessel.

Some features of the metaphor are coincident with those of the building metaphor – the hull corresponds to the foundation, and both possess superstructures. This is not true of Quine's metaphor of the web, though there is the idea of centre and periphery:

> Quine's new model pictures a belief system as a web or net. Beliefs that are most likely to be given up in the face of recalcitrant experience are located at the edges, beliefs less subject to revision are nearer the centre. These latter beliefs are less subject to revision not only because they are further from experience, but also because they are interconnected with more elements in the rest of the system. When experience necessitates some changes in the system, they are usually many ways to revise including changing the meanings of some terms, revising theories or even, Quine hazards, revising logic.[168]

---

[167] Colyer, 2001a, p. 374.
[168] Murphy 1990, p. 8.

It is a metaphor taken up by Rorty and Putnam[169] and finds theological use in Swinburne.[170] It appears that Alston is also describing a web-like structure:

> Perhaps a more reasonable aspiration for the human condition is to have multiple sources of support such that although each can be questioned and none renders any of one's beliefs absolutely certain, they lend support to each other as well as to the beliefs they are invoked to support; so that in the way the whole assemblage fits together we have sufficient reason to take the beliefs to be true … Though each of these considerations can itself be doubted and though no single strand is sufficient to keep the faith secure, when combined into a rope they all together have enough strength to do the job.[171]

Despite Murphy's designation of centre and periphery, the metaphor lacks the progressive layeredness – inherently ordered strata – of the foundational metaphor, and whilst each node may denote a different sort of junction, with varying strength and direction of outward vectors, the overall effect is one of flexibility, fluidity and homogeneity.

All three metaphors therefore are employed theologically, but it is the first two which have most resonance within the Christian tradition. The actual boat of the Galilean fishermen becomes a natural metaphor for the ark of the church. The foundational metaphor of God as the rock, upon which Israel's and the Church's faith is based, is found throughout the Scriptures and in hymnody, furthermore in the New Testament assuming Christological form (1 Cor 3:11, Eph 2:20, Acts 4:11). For Torrance these are not just resonances, they have direct epistemological implications. If the Church's faith is built upon no other rock than God, save at the impudence of idolatry, then the epistemological pole of that faith,[172] our noetic structures, cannot be ultimately built securely upon anything else but God as he is revealed in Jesus Christ, albeit that the mediate nature of the revelation means that those structures assume a transcendentally non-foundational form. Unlike most other thinkers in the field who use the doctrine of creation to explicate the relation of theological and scientific epistemology, Torrance's lynchpin is here Christology.[173]

---

169 Rorty, 1991, p. 26.
170 Swinburne, 1991, p. 174.
171 Alston, 1991, p. 306; see also Osborne, 2006, p. 403.
172 Torrance, 1965, p. 163.
173 Luoma, 2002, p. 7.

There is another metaphor which we encountered in the previous section: the circle, particularly important in the notion of the hermeneutical circle. Logicians have a prior and correct prejudice against circularity, for in their context circularity equates simply to invalidity. However that is not necessarily the case in hermeneutics, where it is impossible to proceed in the linear manner of the logician. In hermeneutics, one can never 'be done' with the text and move on, anymore than one can be done with empirical data. Rather, in the cycle of repeated engagements with the text, in which our own pre-understandings are challenged and reshaped, so our understanding of the text deepens. As we have seen, Barth and Torrance's dogmatics are thus knowingly circular. In Rickman's introduction to the writings of Dilthey, we note a subtle but crucial change to the metaphor:

> This circularity – or perhaps one might call it a spiral approximation towards greater accuracy and knowledge – pervades our whole intellectual life.[174]

To move in a circle is to make no progress, but to go constantly over the same ground – it is epistemological 'groundhog day' – the record is stuck in the groove. To move in a spiral is progressive and refining, and actually no two points on a spiral are the same – the record plays the song. Indeed spirals with a negative pitch not only move forward, they close upon but never achieve a single focus: they are 'convergent' – note Couvalis' 'convergent realism'.[175] Circular reasoning goes nowhere, it is useless and pointless; spiralling reasoning is progressive, providing an ever tightening focus on the object in question, equally as progressive if not as direct as linear reasoning. If the demise of foundationalism cuts off one path – and Feyerabend has shown us that natural science with its 'hermeneutical' to-and-fro between data and theory is on the same bank of the river as theology – then this new path offers a route of possible epistemological progress. McGrath likens the spiral analogy to that of an iterative process:[176]

---

[174] Rickman in introduction to Dilthey, 1976, p. 11; see also for instance Osborne, 2006, p. 83.

[175] Couvalis, 1989, p. 130ff, as a challenge to Feyerabend.

[176] A note on language: 'circular' as we have noted, has entirely negative connotations within the field of logic, which prejudices us against its use. Spiral is a modification which suggests progression, but retains radiality. 'Iterative' and 'feedback' (the latter can be positive or negative) are semantically linear, but bi-directional. This may lead us to favour one term over another. However the differences are essentially semantic in this context: what we are rejecting is any notion of a wholly linear and

> This can be thought of as a process of convergence (=closure) on a region of doctrinal stability, or as ascending a hermeneutical spiral, allowing access to enhanced levels of appreciation and understanding …
>
> The second visual image is that of a spiral – perhaps a spiral staircase. Each completion of the iterative cycle moves us back to the beginning – but also up a level.[177]

This progressive possibility means that we do not have to side with those, including Feyerabend, who have drawn more radical conclusions from the demise of foundationalism. Those conclusions include non-realism (nominalism) and relativism, which Torrance is anxious to avoid. According to McGrath, Torrance is correct to avoid such radicalism: "The rejection of classical foundationalism does not necessarily entail a rejection of realism, or give intellectual legitimisation to any such slide into a freewheeling relativism."[178] Couvalis too contends that voluntarism should have drawn Feyerabend away from relativism, and instead towards a position more like that of Torrance:

> for theorists like Feyerabend, who think ethical considerations should decide the choice of epistemology, this should be a decisive argument for choosing an objectivist but anti-foundationalist epistemology rather than a relativist one.[179]

It is a view shared by the school of Plantinga and Wolterstorff, who recognize that there are those, such as Rorty, for whom the demise of foundationalism implies that the basis of human knowledge is fatally undermined. However, Torrance's fellow Calvinists, Plantinga and Wolterstorff themselves categorically deny this entailment, concluding the relevant passage in the introduction to *Faith and Rationality: Reason and Belief in God* with: "most emphatically these essays do not draw those conclusions."[180] Lindbeck also argues in much the same way, although for him the demise of foundationalism *does* entail a retreat from a full-bodied correspondence theory of truth, to one in which dogmatic statements define the grammar but not the content or meaning of inter-contextual dialogue.[181]

---

unidirective deductive progression – any 'hermeneutic' mode of understanding entails a 'back and forth', which may or may not be progressive and convergent.

[177] McGrath, 2006, p. 195, 202.

[178] McGrath 2002, Vol. 2, p. 33; see also same volume p. 38 and McGrath, 1998, p. 14.

[179] Couvalis, 1989, p. 135.

[180] Plantinga & Wolterstorff, 1983, p. 4.

[181] Lindbeck, 1974, p. 130.

All three of these metaphors – ship, web and spiral – avoid the fatally constricting rigidity of foundationalism, and whilst Couvalis identifies the rejection of such foundationalism as the key to Feyerabend's thought,[182] the three 'alternative' metaphors of epistemological structure allow Torrance and Couvalis (and us) to avoid the anarchy which Feyerabend prescribes as 'medicine'. Non-foundationalism does not render human knowledge shapeless and anarchically unstructured, forcing us to resort to clinging onto driftwood in an open sea of relativism – it simply points us towards alternatives. However the question remains of how theology is to speak of its object – God – scientifically, when the data is only known mediately, and when the sovereign subjectivity of that object over us must always be preserved.

Conclusion

Feyerabend's challenge to foundational notions within natural science are at their genesis unlikely to worry theological scientists, for they are rooted in criticisms of logical positivist notions of theory and observation, a philosophy which is implacably hostile to theological and metaphysical notions in any case. In fact as we uncover the hermeneutical circles which lie within all acts of human understanding, so the increased consonance of scientific method over the fields of natural and theological science is revealed, and the Enlightenment privileges accrued to the former are radically scrutinised. Feyerabend's assertion that all facts are ideational, and therefore accompanied by acts of human interpretation – conscious or un-conscious – develops into a more general critique of foundational schema, though he is reluctant to commit himself to any rigorous form of relativism which runs the risk of self-refutation. However, with the aid of the incommensurability thesis, having ditched the strong realism of his earlier philosophy, concrete notions of scientific method, rationality, objectivity, even truth, are radically deconstructed and replaced by a voluntarism that elevates ethical choice – again conscious or unconscious, humanitarian or oppressive – and his mid-term prescription for science and epistemology is anarchy.

[182] See Couvalis, 1989, in particular p. vii: "He [Feyerabend] argues that the task of trying to find foundations for knowledge should be abandoned because there are no such foundations to be found." Also Feyerabend's 1961 paper: *Knowledge without foundations*.

Whilst Torrance therefore sees some advantages in the opening gambit of Feyerabend's challenge, he rejects Feyerabend's more radical conclusions. Feyerabend's more heady notions of theoretical pluralism and incommensurability, and what follows from them, depend in turn upon unwarrantedly strongly theory-determined observations. Torrance rejects this theory-determination (as do others) insisting that whilst theory creation cannot be made any less of a speculative activity, it is nonetheless the relentless critical search for a unitary depiction of reality, ultimately aiming to dispense with random or unjustified social components.

For Torrance, Feyerabend's deconstruction of science as a concept is unsuccessful, though he would concur that inherited components of its meaning may need to be ditched or rethought. For Torrance there is the possibility of some sort of foundational representation of knowledge, employing a fluid axiomatics, though if viewed through the lens of Enlightenment concerns it may bear the appearance of circularity, or at least an iterative and fluid spiral. Reductive components of Enlightenment foundationalism, the primacy of empirical data and deductive techniques, are refashioned by Torrance into a new shape which steers a moderate path between inherited structures and Feyerabend's radical critique. If theology is to give an account of the being of God, known in the event of his revelation in Jesus Christ, then that science will have a fideistic quality, for it is mediated, even for those eye-witnesses to the event, through the prism of human belief or un-belief.

That prism also affects the extent to which Torrance is able to win public support for attempts to give warrant to realist assumptions regarding issues of correspondence and coherence, because even here the ground, ground that is often considered to be purely philosophical, is staked out with theological concerns in order to respect the graceful sovereignty of God. Furthermore such justifications are also dogged by the non-scientific nature of religious language.

The task of the theological scientist, who may also be the minister of the Word, cannot therefore be apologetic in the sense of telling, of representing in logical form, for what would thereby be told would be the assertion of a narrative to which post-modernism can grant no privilege at the risk of caprice, and must therefore, remembering that the gift of faith belongs to God's prerogative, content itself with the apologetic task of showing – of coherence, unity and depth.

CONCLUSION

Though a quite unforgivable historical cliché, we begin our conclusion by saying that the world changed in 1914. For underneath most clichés is some foundation in truth. We admit the Euro-centric outlook of the cliché, which locates the change in a particular military event, the outbreak of the first world war. Any statement regarding the clash of the great European imperial powers, the beginning of the final decisive waning of those powers, the subsequent changes to the social and political order, the sheer scale and futility of the waste of human life, is bound to degenerate into further cliché. However there are other reasons, more closely associated with our own themes to suggest that the world was undergoing profound changes.

Einstein's special and general theories of relativity were still new, controversial and awaiting empirical verification. Quantum theory had recently begun to revolutionize particle physics.[1] Russell's paradox, so decisive a blow to Frege's project, was a decade in the past, and his own *Principia,* jointly written with Whitehead had been published the year before. Gödel's theorem, the shattering blow to mathematical formalism, was still nearly two decades away. Wittgenstein's *Tractatus* was in gestation, as were many of the ideas of the Vienna circle, its institution being still nearly a decade away, though Schlick's *Allgemeine Erkenntnislehre* was written during the war. In theology, the historical critical school of liberal Protestantism had just passed the apogee of its influence, the 4th August 'manifesto of the intellectuals' marking a decisive turning point in its credibility.

One theologian who recoiled in horror at the acquiescence of his teachers to the military purposes of Kaiser Wilhelm II, was Karl Barth. It was precisely at this point that Barth understood that theology must be unreservedly founded on God's revelation. He soon abandoned the *Christian Dogmatics,* embarked on his cathartic exploration of Anselm's theology, wrote and published the landmark *Commentary on the Romans* (first edition 1919), and began the magisterial *Church*

[1] A more piecemeal process then relativity theory, not least in its interpretation. Paul (1982, p. 72) cites the birthdate as Einstein's 1905 paper: "His explanation of the photoelectric effect was eventually recognised as the birth of quantum theory with Planck's innovation regarded as its conception."

Dogmatics which was to dominate Protestant theology in the twentieth century. In other more distant fields thinkers such as Stravinsky, Picasso, Weber, Proust, Joyce, Freud, were changing their disciplines for ever, and the fifty year old ideas of Marx were about to be incarnated in the Soviet Revolution.

Far away on 30th August 1913 in China at a remote missionary station Thomas Forsyth Torrance was born. Ten years later in Vienna, the place that had for so long been the centre of European culture, Paul Feyerabend was born. In 1914 the 23 year old Hungarian chemist Michael Polanyi began to assimilate the great and terrible events which surrounded him. The Enlightenment dream of rational unitary progress in all areas of knowledge, founded upon the ethical solidarity of the brotherhood of all men – it was a brotherhood, sisterhood was still a way off – lay in tatters, and a more fragmented, less self-confident world emerged, twenty years later to be plunged into an equally deep and damaging crisis. Spengler's *Decline of the West* (1918) and Bury's *Idea of Progress* (1920) seemed to encapsulate the thoughts of the post-war generation.

Logical positivism, however flawed it may appear with the aid of hindsight, must have seemed to many like some sort of haven from the storms which raged around them, a place where on the ground of protocol sentences, however unlikely now that base might seem, something solid might yet be salvaged.

In Karl Barth's view, within the past (i.e. the nineteenth) century, Protestant theology had taken a decisive turn towards anthropology.[2] It had abandoned consideration of its proper and unique object, and had substituted in its place what was essentially a study of man,[3] seeking to establish itself on the common ground of human reason, which had assumed so central a place since Kant's 'philosophical Copernican revolution.'[4] Barth, particularly in his definite rejection of natural theology, sought to found theology entirely on the self-revelation of God in Jesus Christ, an event which he believed was self-interpreting and self-defining. Torrance, his English interpreter, sought to further Barthian theology, and also to set out the implications for theology of

[2] Pannenberg, 1976, p. 265.

[3] Torrance, 1988, p. 151.

[4] Popper, 1979, p. 92 – the analogy is apt – but it is often unnoticed that cosmology went through another paradigm shift: the relativistic. This extended analogy would provide philosophy with a way past the old Kantian antimonies to the kind of Kierkegaardian reassessment which Torrance proposes.

the developments which we have outlined in the natural sciences, particularly the advent of quantum and relativity theory.

Paul Feyerabend was himself initially attracted to the safe haven of logical positivism, as a student ostentatiously asserting the pre-eminence of the natural sciences in the field of human knowledge. Then being converted, as he no doubt was, despite later recantations, to Popper's falsificationism, he found an alternative rationalist haven for his thoughts. However over a period of twenty years, there was a rather fragmented but nonetheless decisively anarchic disassembling of his thought. To speak of Feyerabend *arriving* at the position of *Against Method* is to impose a rationalist progressive framework on his epistemological anarchy. Rather what he 'arrived at' was a point at which various themes and dialogues are found in perpetual motion. Furthermore in subsequent publications, he was never afraid to 'circle' around a topic contrapuntally and often ironically, to the frustration and sometimes fury of his commentators and opponents.[5]

* * *

The conclusion I offer is to describe the outcomes of the constructed encounter between Torrance and Feyerabend as follows. The constructed encounter between Torrance and Feyerabend is best envisaged as Feyerabend posing an anarchic challenge test to the coherence of Torrance's project of theological science. Both Feyerabend and Torrance were writing in a context characterised by naturalistic and scientistic conceptions of 'the scientific', which they both felt warranted critique. Torrance wished to redefine the content of the predicate 'scientific', refining it to the form of the epistemic virtue of objectivity understood as 'faithfulness to object,' superseding an impersonal objectivity which had hardened into objectivism. Although this was primarily no doubt intended to rehabilitate the scientific status of theology in the style of Karl Barth, in order to avoid the accusation that this refinement was simply an ad hoc justification of that status – unlike Barth, who was hostile to any imbedded notions of equivalence between the theological and natural sciences – Torrance sought to capitalize on recent changes within natural scientific paradigms to show that here too a new climate was emerging which was forcing natural 'science' itself to reassess its essential properties.

---

[5] For instance in *Farewell to Reason* and *Three Dialogues on Knowledge*.

Whereas other theologians have more commonly investigated scientific paradigms in order to reflect on the effect of their content for specific Christian doctrines, Torrance examines such developments as relativity theory, quantum theory and the Gödelian theorems in order to assess the philosophical meta-conditions which they entail or presuppose, thus gauging the 'climate' in which theology will have to function. In order to achieve this, Torrance specifies various similarities and differences which we have examined. However beyond examining these individually we must also ask whether the former are substantive enough to offer Torrance's conception of science a coherent core, or, not only whether the differences constitute special pleadings for theology, but furthermore whether those differences are strong enough to have the capacity to undermine, as Feyerabend would no doubt maintain, any integrity to the conception 'science' afforded by the similarities.

The coherency of theological science is therefore sharply challenge-tested by Paul Feyerabend. Whilst accepting that positivistic versions of science will not do, Feyerabend offers a more radical critique of 'the scientific', asking whether the category is helpful or coherent at all. Starting with a critique of the observation / theory divide in positivistic understandings of science, Feyerabend, by way of the ideas of counter-inductivism, theoretical pluralism and incommensurability, gradually, and it has to be said somewhat chaotically and fragmentally – though that is part of the point – deconstructs the whole notion of science, and by examining the sociological factors within human knowledge, seeks to show that it is irrational and human forces which decisively shape what is classified as good science. Noting that this classification has often, at least in his time if not in the past, served to defend the vested interests of the powerful and wealthy against more marginalised traditions, he operates with a hermeneutic of suspicion towards the whole notion of science.

Feyerabend's challenge to Torrance is therefore to assert that Torrance has been too timid, trying again to ingratiate theology to an indelibly imperialistic scientific establishment. It is ironic that this is Barth's accusation against liberal Protestant theology, an accusation which Torrance supports. According to Feyerabend theologians should not be content with reshaping or readjusting our notion of the scientific, they should be, alongside other marginalised traditions, demanding that society abandon such a notion, in favour of an anarchic 'theory' of knowledge which simply promotes human flourishing. Theology should stop being timid and start standing up for its rights. Epistemic criteria and virtues have become hijacked and employed for science's own ends, and they must now be discarded, and ethical and political considerations put in their place. Freedom should become science's goal, and truth – along with notions of objectivity and rationality – are better dispensed with, for they simply become idols distracting us from our primary purpose. The criteria which science sets for good method – objectivity, rationality and truth – are too complex, too intertwined with the vested interests of the powerful to be disentangled – rather they must be abandoned and more humane criteria put in their place. Certainly Feyerabend admits that what we classify as natural science may still do well by those new criteria, but it will not be in a position to foist itself upon its less-powerful rivals.

As we shall note it is theological concerns which underpin Torrance's response, and certainly they are present here. When Feyerabend seeks to persuade us to abandon the search for truth in preference for the search for freedom, his liberal and optimistic outlook on human nature assumes that a degree of human solidarity will facilitate that search being at least moderately successful; Torrance operating with a Calvinist anthropology[6] is wary of such an assumption and estimates that self-seeking and self-projection are far more likely results. Whatever the roots of their respective anthropological outlooks, the optimistic and pessimistic elements contained therein, are determined by wider concerns. Feyerabend is optimistic about the judgements of society as a whole, but rather pessimistic about those who belong to the scientific class within that society. Conversely Torrance is pessimistic about human nature in general, but rather optimistic about the abilities of the class of theological scientists to be disciplined and self-critical enough to transcend Feyerabend's vicious circles. However ultimately Feyerabend can offer us no convincing reason why we should accept his conflation of the epistemological aims of science with the social and ethical conditions under which it operates, and Torrance's (and others') view that one should pursue the 'truth that sets you free' seems an adequate rejoinder. However it does require that such truth-seekers possess an adequate self-critical faculty, so that their own desires,

[6] According to Calvin, we are "alienated from God in the whole of their mental system", Torrance, 1949, p. 119 – though also "the human mind … is still adorned and invested with admirable gifts from its creator," (Calvin, *Institutes,* II/2/15), should not be overlooked.

conscious or sub-conscious are not unwittingly projected from within that 'truth'.

When Feyerabend appeals to theology for support, he does so as to a fellow outsider and therefore a possible present ally of his radicalism. Theologians such as Bultmann, but also Torrance, are, according to Feyerabend, ultimately collaborators, who, whilst they might rationalise their position as moderates trying to reconcile theology and natural science, are liable eventually to emerge as traitors. Rather like the 'Californian fundamentalists'[7] (Feyerabend's example) they should join the resistance, in what will become a post-modern reaction to the dominance of the Western natural scientific meta-narrative within modernism.

Stimulating and sometimes transfixing as his rhetoric might be, I submit that ultimately Feyerabend's challenge is a failure.

His rhetoric[8] is just that – witty and stimulating, but not backed up with rigorous arguments. His historical examples by their paucity and one-sidedness ironically exhibit a Popperian falsificationism of which he is such a vociferous opponent, an irony Newton-Smith has observed.[9]

His style of collage, means that he fails to assemble the necessary parts of his argument in proper alignment – for instance, his incommensurability thesis is developed at a time when he is an exponent of super-realism, yet when he wishes to use the thesis in his relativist and voluntarist phase, his super-realism has been long abandoned, and incommensurability becomes a weak idea without it. Crucially, his construction of straw men, and dichotomous approach to questions – such as the theory infectedness of observation – means that he fails to engage with or respond to mediating positions such as Torrance and Polanyi. When Feyerabend says, for instance:

> What we find when living, experimenting, doing research is therefore not a single scenario called 'the world' or 'being' or 'reality' but a variety of responses, each of them constituting a special reality for those who have called it forth. This is relativism because the type of reality encountered depends on the approach taken[10]

– we are entitled to respond that what he describes is in fact not relativism at all – at least in any but the weakest form of the term. It is certainly

[7] Feyerabend, KSR, p. 187.
[8] Finocchiaro, 1975.
[9] Newton-Smith, 1990, p. 134.
[10] Feyerabend, AM, p. 270.

not a relativism strong enough to send critical realists running for cover. When there is a spectrum of alternatives (e.g. objectivity / subjectivity), providing counter-examples of one extreme of that continuum does not entail the truth of its opposite – a "monumental *non sequitur*"[11] – rather a more nuanced approach must be sought.

Whilst Feyerabend's rhetoric might accuse those like Polanyi and Torrance of timidity, and has the glamour of an epistemic revolutionary and freedom-fighter – appeasement, negotiation and syncretism are sometimes what the situation demands. Most decisively his anarchic project of 'knowledge without foundations', which so ably deconstructs Enlightenment foundationalism, has been shown not to entail any of the more radical conclusions which Feyerabend demands, by Torrance and by others. In particular, non-foundationalism does not mean that the edifice of scientific knowledge collapses in the way that Feyerabend assumes, but rather by a bit of shoring up and underpinning can retain a humbled and fluid structural integrity. In particular Torrance's notions of a critical realism, a new objectivity, a posterior and variegated rationality, an immanent scientific method, fluid axiomatics, a spiralling, iterative and hermeneutical progress, all emerge as competent alternatives to their older, more naïve, universal and simplistic forebears, and resilient to Feyerabend's critique, therefore further developing the consonances between our present scientific paradigms and theological science. Hermeneutics assumes a ubiquitous role in acts of human understanding, not as the poor relation of deduction and induction, but as a necessary and valuable component of all the sciences.

Another decisive weakness in Feyerabend's approach is that once he has adopted the unit of tradition, he fails to examine what constitutes such an item and therefore assumes, contrary to the historical evidence, that traditions can only be understood as homogeneous, isolated and essentially hostile sociological entities – certainly sometimes they can be – but by making such an assumption universal, he has committed *petitio principii* by implicitly including in his 'definition' what he wishes to prove, namely that a particular tradition – Western natural science – has consistently acted with hostile intent towards others. Certainly that may have been true, and continues to be so within the on-going and rather sterile and ill-informed debates between fundamentalists and materialists, so that the impression continues to

[11] Laudan, 1996, p. 104.

be given that the conflict model best expresses the interaction of natural science with Christian theology. However Torrance convincingly shows us that historically this has not always been the case, and that neither is such a view necessary on more abstract epistemic grounds. At times natural science has determined that the climate will be one of mutual hostility, and there are those today in the 'New Atheists'[12] mould who would wish that it remains so – but they are living behind the times, and the scientistic, positivistic and materialistic climate on which they are parasitic has become more temperate for theological notions. Feyerabend has not balanced his hermeneutic of suspicion with a corresponding hermeneutic of obedience, so that the tool of suspicion asserts itself as a world-view, meaning in turn, that traditions as they engage with each other and as he describes them, are unable to speak clearly and on their own terms.

My interest in both Torrance and in particular Feyerabend is partly based on their status as pioneers. Feyerabend's standing as a philosopher of science was not enhanced by his radicalism, which was mostly judged by that community to have been immoderate and unwarranted. Although he was hurt by that judgement, simultaneously his ideas were taken up by varied external groups of people, particularly within the social sciences, sheltering under what came to be known as the post-modern and anti-technological umbrella. Whilst it is beyond our scope to attempt any assessment of all that followed, it does force us to question where we might place Torrance in more recent developments – a question which, as we have seen, is also relevant to current assessments of Barth. Like Barth and Feyerabend, Torrance clearly believes that modernism had forced natural (and theological) science into too tight a philosophical strait-jacket. However Torrance and Barth cannot, with their emphasis on objectivity and realism, be classed with theological post-modernism. Neither, despite his Barthian emphasis on the prevenience of revelation can Torrance be understood as simply attempting nostalgically to reinvigorate the easy certainties of pre-modern patterns of thought. Other considerations aside, his constant engagement as a theologian with cutting edge natural science – his message to many of his theological contemporaries can be best expressed as 'keep up!' – eliminates that interpretation. Rather he stands in a mediate position between the Scylla modernism and the Charybdis

[12] Dawkins, Harris, Dennett, Grayling, Hitchens, Toynbee, Amis, McEwan in Beattie, 2007.

post-modernism, though on such a fideistic platform, embedded within such a particular theological schema, that it is doubtful his approach will be widely convincing.

I believe therefore that Torrance's conception of science survives Feyerabend's challenge test, but not without sustaining real damage. Although Feyerabend's case for 'indirect refutation' in physical science – alleging that in certain cases, developments in one theory can only be achieved by advancements in another – was weak, nonetheless it is incontestable that, in the hermeneutical sciences, work in one field or school can highlight weaknesses in another, as here is the case.

1. Firstly, the acceptability or otherwise of the fideism inherent within Torrance's approach will no doubt depend to a large extent on the prejudices of his readers. However what is not possible, is to abstract from Torrance a prior non-theological epistemology which shorn of its theological commitments can itself rescue science from Feyerabend's challenge. Certainly it is possible to propose non-theological responses to Feyerabend – and very convincing ones – but Torrance's approach is distinguished from these, being theological 'all the way down':
   - with respect to foundationalism, 'Christ is made the sure foundation' is a statement with precise epistemological and scientific import;
   - in terms of objectivity, the overwhelming subjectivity of God, given kenotically for us as object in the incarnation, requires that objectivity be understood as faithfulness to object, with faithfulness retaining all its theological resonances;
   - in terms of rationality, the *Logos* of God as differentiated from, but the source and sustainer of, the *logos* of creation;
   - in terms of reality, the ultimate reality of God, made known in the event of revelation, and there condescending to be given analogical semantic reference, undergirds a scientific realism which demands that scientific statements iteratively and verisimilitudinously achieve conformity with the nature of created reality within natural science. Dummett's argument[13] that realism is *only* tenable on a theistic basis – Moore, like Torrance, substitutes Christian for theistic at this point – is the basis of the argument here.[14]

[13] Dummett, 1978, p. xxxix, also p. 369 – "certainly theism helps us out with some notoriously testy problems".

[14] Moore, 2003, p. 7.

Theology underpins Torrance's epistemology because theological imperatives are *necessarily and irreducibly* foundational, and without them Torrance deliberately leaves himself open to Feyerabend's attack: *epistemologically, Torrance is determined to live by grace alone*. Furthermore though not the focus of this book, we have noted that Torrance's theological schema is a particular one, and given that the content of the notion 'scientific' is embedded within it, other schemas will play out in different ways.

The search for non-theological epistemological bedrock represents for Torrance our desire for ultimate and eternal certainty, it is the search for an idol, an attempt to escape what we actually are – finite, created and dependent – it is an attempt to be gods. It is an attempt to transcend our proper creaturely objectivity for an Archimedean objectivism.[15] Therefore the Christian can only renounce it. Our theological identity as creatures[16] is an inescapable feature of epistemology and reality.

To say therefore that Torrance's epistemology is theological is not simply to state the obvious for a Christian theologian – but to draw a contrast with those such as Pannenberg, more willing to play the game of the enlightenment thought experiment. Torrance limits himself to offering a coherent, but nonetheless converting account of Christian faith, so that his task is not apologetic to the extent that every move can be defended in theologically neutral or logical terms – for whilst faith is by no means contrasted with reason, it remains the necessarily extra-logical gift of God, a gift which is itself reason-making: "faith here simply means fidelity of the human reason to what is actually there in the encounter."[17] The existence of God is the assumption on which Torrance, the dogmatician, proceeds, just as the natural scientist proceeds on the assumption of the apprehendability and intelligibility of the natural order.[18]

For Torrance it is only by utilising a strong and covenantal[19] doctrine of creation that this assumption of a natural order can be underwritten. Creation, being not-God, is therefore not a simple partaker in the divine rationality (Torrance distinguishes here the mistaken Thomist *analogia entis* in which theology utilizes whatever

15 Griffith-Dickson, 1995, p. 302 "the foundational crime of pure reason".
16 Brown, 1955, p. 172.
17 Torrance, 1969, p. 325.
18 Colyer, 2001a, p. 30.
19 Torrance, 1990, p. 93.

"metaphorical resources are at hand," from the more proper Barthian 'analogy of faith' "in which our knowledge of empirical worldly reality is subsumed through revelation within a divinely determined meaning structure"[20]) but nonetheless is graciously gifted its own derivative and created rationality (*logos*),[21] which reflects and mirrors that of its creator,[22] though it must never be confused with it.[23] However, Torrance's high Christology, in turn implies that the divine rationality (*Logos*) is once made present in creation, in the event of the incarnation, where our knowledge of that *Logos* must be mediated through the *lalia* of fleshy speech. Though that *Logos* is thereby rendered mediate in that all human utterances, especially those which we must access via a written historical record, need to be 'heard' – there can be no mystical[24] means of circumventing those *lalia*.[25]

Natural science is therefore enabled to investigate the natural world, free of any latent divinisation, by methods and with tools appropriate to that object. Thereby it will discover a world which is well-ordered, beautiful and regulated, but these are natural science's own fideistic commitments, rather than its findings (though confirmed *a posteriori*), and this 'inherent embedded rationality'[26] cannot speak more directly of the nature of its creator: the search for 'elemental forms' is predicated upon the assumption that such forms exist, and are constituted in the natural *order*. The existence of such a rationality – understood as an integral feature of creation, rather than that which we impose upon it – corresponds to Hacking's contrast between 'inherent structuredness' and 'nominalism' – Torrance is a theologically motivated and underwritten inherent structuralist.[27]

Torrance also proposes that properly understood this relation of creator and creation explicates some of the metaphysical puzzles and epistemological connections that worry philosophers.[28] These

[20] Patterson, 1999, p. 13.
[21] Torrance, 1995, p. 320.
[22] Torrance, 1995, p. 220.
[23] "God functions both as a warrant and as an explanation for the rationality of the world" – Smedes, 2004, p. 95.
[24] See reply to Richardson in Colyer, 2001.
[25] Torrance, 1965, p. 141; Torrance, 1971, p. 139.
[26] Torrance, 1971, p. 42.
[27] Hacking, 2000, p. 83.
[28] Also Moore, 2003, p. 7.

connections can only be properly seen in the light of that theological relation, so that viewed from the standpoint of non-faith, they appear simply as opaque and unfathomable, like vortices, the currents of which pull the thinker down towards a circular relativism not unlike Feyerabend's. As Swinburne maintains, the existence of a creator is the hypothesis which most *simply* extricates us from such quagmires.[29] Polkinghorne also concurs that "those who wish to understand more deeply this remarkable property of the rational transparency of our world, should, in my opinion, first look to a theology of creation rather than to a philosophy of knowledge."[30]

Whilst Torrance's epistemology is theological 'all the way down', we have also noted that 'on the way up', we set ourselves a formidable task – the "Christian reconstruction of the foundations of culture."[31] Whilst the scale of the task should not in itself dissuade us from the true path, one can understand why Feyerabend's more immediately utilitarian course might entice those more easily intimidated.

2. Secondly, Feyerabend's hermeneutic of suspicion seeks to deconstruct the cosy and self-reinforcing vicious circles of any discipline which claims for itself the label scientific. It is certainly the case that in theology, perhaps more than in any other science, the history of ideas and doctrines has been laden with political interests, which were given huge sociological potency in (for instance) the terms 'orthodoxy' and 'heresy'. To the modern scientific observer the normative councils of Ephesus, Chalcedon and Nicea, seem overwhelmingly burdened with political intrigue, narrated for instance in Walter Bauer's *Orthodoxy and Heresy*. Only by a far greater integration of dogmatic theology (as historical theology) and church history than is attempted by Torrance,[32] can Feyerabend's hermeneutic of suspicion be rebutted. MGrath also notes[33] that Torrance's personal objective knowledge opens up fresh possibilities for a renewed integration of theology and spirituality, though again there is little actual progress in this direction within Torrance's corpus.[34] Torrance's holism demands such a unified project, but it is understandable that given the scale involved, he can only point in

[29] Swinburne, 1991, p. 287.
[30] Polkinghorne, 1998, p. 109.
[31] Torrance, 1965, p. 271.
[32] Pannenberg (1976, p. 372) lays the blame squarely at the feet of Barth for this.
[33] McGrath, 2002, p. 202.
[34] Hing-Kau, 1996, p. 209.

the direction of travel. Torrance's largely ideational account of theological science reflects a more widespread division between these disciplines, which particular theologies, such as liberation theology, seek to overcome. Although Feyerabend's praise for liberation theology is purely on the criterion of *praxis*, it is surely no co-incidence that it seeks to expose relations of power and inequality which have informed, consciously or unconsciously the formation of Christian doctrine. Here Feyerabend does point to further work which theological science needs to do, and his critique gains real purchase.

3. Thirdly, Feyerabend's Kuhnian shift towards historicalism highlights the fact that in trying to exposit a 'classic' Christian tradition (analogous to 'good science'), Torrance commits himself to various 'distillations' of historical theology. These distillations are at times both rather idiosyncratic,[35] and it is also unclear as to whether such distillations are *describing* (something which is self-evidently 'good science', and therefore which our criterion of 'classic'[36] must conform with) or *prescribing* (something which deserves to be part of the classic tradition, because it has met some prior criteria of 'good science'). So in assuming that ideas will eventually 'win out' in theological science, he fails again to address the real concern that political and ecclesial motivations of power and religious enthusiasm have determined the outcome of doctrinal disputations – bridges towards church history must be built as soundly as those towards Biblical studies and natural science.
4. Fourthly, by his utilisation of a contextual theory of language operative across all the sciences, Feyerabend highlights the fact that any increased through-flow of traffic on these inter-connecting bridges will entail using a theological language which operates in a variety of contexts, including those which militate against scientific transparency and precision. Theology cannot mould its language into the sort of artificial state which natural science attempts, rather it must operate with a language naturalised and formulated within all branches of theology and church life. Torrance does not account adequately for this complicating and enriching factor.
5. Fifthly, whilst Torrance succeeds in showing how theology and natural science *might* co-exist and cohere within a unitary epistemological framework, he does not succeed in showing that natural

35 Luoma, 2002, p. 22.
36 McGrath, 2001, p. 42.

science has or will pay any particular attention to or benefit from the findings of theology. Whilst the conclusions he draws from developments within natural science and mathematics may inform a 'climate' in which theology may flourish, it is not and cannot be, that any of these developments are *necessary* results for theology, and when he presses hard on changes in quantum mechanics, relativity theory, and the philosophy of mathematics, there is a danger that he is extending these results to beyond their domain of validity. Subsequently his almost totally uncritical appraisal of recent developments in natural science, particularly within modern physics, coupled to his contingent yet deep alliance between theology and natural science, means that theology appears more or less to have married the present physical paradigm: precisely the charge he levels at earlier attempts to force theology and natural science to the altar. The absence of any sense of tension between natural and theological science, implies that theological science *has* conformed to natural science once again, and there is no real attempt to show how theology might inform or transform the current paradigms of natural science. This 'all good news' appraisal of contemporary developments in natural science, may lead us to suspect with Feyerabend, that Torrance is simply ingratiating himself and unwittingly collaborating with Western science's imperialistic habit.

6. Finally, Torrance's confessionalism hampers theology's attempts to move beyond initial fideistic commitments and positively engage with wider sections of the academy. Whilst Torrance's Feyerabendian point is well made – that theology can only undertake such engagements ensuring at all points that nothing vital to its integrity has been decisively yielded to natural science – such encounters, to use Feyerabend's language, must be 'open' rather than 'closed' – there is therefore an unavoidable pay off between 'general validity' and theological integrity, which will mean that theology is likely simply to be ignored. However whilst the sovereignty, Lordship and utter subjectivity of God cannot be undermined or compromised, it is not clear whether those qualities, need to be so jealously guarded by theological science. The Christian may have to play the enlightenment game with fingers crossed, but refusing to take the field lest the sovereignty of revelation be comprised, seems haughty and condescending in exactly the way of incarnation is not. God's aseitic subjectivity should lead us rather to conclude that God can handle himself in this regard, and that if evangelically advantageous, there is no reason why theologians cannot participate in

imaginative theological exercises – 'get their hands dirtied' – analogous to natural scientific thought experiments and modelling (or in mathematics, *reductio ad absurdum*), which whilst not committing the scientist to alien prejudices, allows him to immerse himself (as Feyerabend requires if the encounters are to be honest and open) in the other's cognitive world – such 'heterocentricity' itself being a vital component of Torrance's objectivity. Pannenberg's project and the search for the historical Jesus might be defended upon these lines, though admitting, for instance, that if ensnared within the positivistic framework of certain schools of historical science, good scientific procedure demands that theology should not simply conform, but confront the prejudices which have thus trapped it – historically a caution which it has not always properly maintained, and the mistake which Barth and Torrance react against.

The distinction between a theology which lives with natural science, and one which marries a particular set of paradigms, is always going to be difficult, for a theologian like Torrance who draws them so closely together, to maintain. He is always liable to be seen as outlining a relationship of necessity which in fact his own Barthian approach proscribes, and which Barth himself eschews on the grounds that the weather was too inclement.

* * *

Despite these weaknesses, Torrance's approach is a more robust and well-measured one than Feyerabend's. The dynamic and contrapuntal nature of Feyerabend's exposition means that he is more likely simply to state a particular circularity, than thoroughly work through its implications, including the possibilities we have outlined for convergence within that circularity. So, for instance, he demonstrates the theory-ladenness of observation, which while devastating for logical positivism, is old news for those involved in the hermeneutical sciences. As Torrance reminds us, once established, the theory-ladenness of observation must be examined to see whether the circularity is inherently vicious, or whether we might yet make progress – whether, crucially to critical realists, it is merely theory-laden or inescapably theory-determined. Feyerabend does not see the importance of this distinction, and passes from one to the other, assuming that the first somehow implies the second.

The same is true of Feyerabend's employment of the Sapir-Whorf thesis. In the weak form of the thesis which simply says that languages are shapers as well as describers of reality, the critical realist is not faced with a crippling rebuttal. The strong linguistic relativist has to prove that the circularity is vicious, that the shaping of 'reality' by language is so impervious to reflection that it robs language of its ability to represent and refer. The contextual theory of meaning implies quite rightly that meaning is implicitly conjoined to use, but that does not entail that words are deprived of their ability cogently to refer to realities beyond language. These relations can be understood as the relations between that which is focal and that which is subsidiary (Polanyi), that which is master and that which is slave, that which we suspect and that to which we submit. Too often Feyerabend simply states that we are not making as much progress as we might have hoped, without actually proving that we have got nowhere – caution is one prescription, anarchy quite another.

At the outset we noted that in order to make Torrance and Feyerabend face each other, we had to remove them from their original context, and whilst we heed their warnings about attempting an 'objective' exchange, by surrounding them with contemporary concerns, we have enabled a dialogue of sorts to ensue. Today logical positivism is no longer taken seriously, and Popper is generally acknowledged to have forced science into rather too tight a falsificationist strait-jacket. Therefore Feyerabend's rhetoric is liable to look like posing and shadow boxing. Ultimately it is Feyerabend's edifice which is more like a Maginot line than Torrance's, bristling with rhetorical artillery on one side, but in fact quite harmless on the other.

* * *

Whilst we might have succeeded in showing the potentialities for fruitful dialogue present within Torrance's theological science, what is lacking is little evidence beyond Maxwell (and one or two other favourites) that natural scientists *have* understood and been either influenced by developments in theological science or even that they have appreciated a change of climate more tolerant to faith commitments, and that we must admit is simply because generally they have not.[37] The danger of

[37] Luoma, 2002, p. 26 – furthermore Luoma later points out (p. 107) that Torrance makes a disastrous omission in his assessment of Maxwell, in that Maxwell's theories are still dependent on the existence of the ether – it is only in dispensing of this feature that field theory achieves the unitary framework of space-time which Torrance desires.

Torrance's confessionalism, is not so much illogicality or capriciousness, but being simply ignored by those who assume that such activities are for those 'who like that sort of thing'. The example of 'cross fertilisation' he gives at the conclusion to *The Ground and Grammar of Theology*[38] – interesting though it is – perichoretic and onto-relational notions within Fritjof Capra's work in particle theory and quantum mechanics, developed further by Christopher Kaiser[39] – is hardly terribly substantive.[40]

Natural scientists, even those relatively few with an active Christian faith, have had little time or need to learn substantive lessons from theology, regarding 'onto-relational' thought or any other part of the theological fabric.[41] Though it is also true to say that natural scientists tend generally to be a pragmatic tribe, and few of them seem to expend much energy on the problems which exercise philosophers of science. They may not have listened to Torrance, but neither have they listened much to Popper, Feyerabend or Van Fraassen.

Moreover as Colyer points out,[42] within the world of theology, whilst there is little doubt in the "growing consensus that Thomas F. Torrance is one of the premier theologians in the second half of the twentieth century," "his work has not had the impact one might expect."[43] The reasons for this he goes on to suggest are a style "dense to the point of obscurity," and the "somewhat unorganised character of Torrance's publications."[44] He laments that Torrance despite his prodigious output, never wrote a projected three volume systematic theology which would have given his theological vision an 'overarching architectonic.' His vision of theology as a refined and rigorous science "is quite frankly incongruent with much of the theological world today."[45] Thus his theology is "repeatedly misunderstood, and his work has not influenced the church and the theological community as significantly as it should."[46]

---

[38] p. 175–7.

[39] "Here we have an instance where Christian theology in its rigorous, scientific form can be of real help even to natural science," p. 176.

[40] Norris, 1984, p. 466.

[41] N. Lash, *Observation, revelation and the posterity of Noah,* in Russell et al., 1988, p. 204.

[42] Also see Marley, 1992, p. 19.

[43] Colyer, 2001a, p. 11 & 15.

[44] Colyer, 2001a, p. 16.

[45] Colyer, 2001a, p. 16.

[46] Colyer, 2001a, p. 17.

Torrance's unitary framework must therefore be regarded more as a defence for theology against positivistic conceptions of science, than a constructive pattern which has been widely influential within the natural scientific or theological community.

* * *

Few theologians make the transition to be natural scientists, and though not large in number, it is certainly more common for professional scientists to become theologians (e.g. McGrath, Peacocke, Polkinghorne). Whilst Torrance[47] is not in that latter category, he is nonetheless amongst the 'life-long' theologians, more acutely aware of the contents of natural science, its methodology and philosophical background.[48] This conversation is only possible because Torrance's work on scientific methodology, natural and theological, is substantive. Too often theologians (and others) operate with or assume an outmoded philosophy of science – or equally likely get carried away with notions of relativity, indeterminacy, and social construction. In his writing on scientific methodology, it is therefore profitable, not simply to compare Torrance's work with that of his fellow theologians, but also with philosophers of science, especially contrasting ones such as Feyerabend who can really challenge Torrance on this ground. In this book I have compared two conceptions of *science,* albeit in one case a conception immersed within a certain theology, and in the other, one which is fundamentally deconstructive.

Whilst Torrance was writing to a world which still held a rather unquestioning respect for the epistemological authority of natural science, it is more interesting forty years later to ask whether that project still has integrity, subsequent to post-modernism's increasingly suspicious view of meta-narratives, including the natural scientific. Feyerabend offers himself as a convenient standard bearer, contemporary with Torrance, for that post-modern critique.

---

[47] Polkinghorne (1998, p. 81) also laments the number of theologians willing and able to seriously engage with science, but notably gives priority to Torrance as one of trio of 'honourable exceptions' (the other two being Moltmann and Pannenberg). His only negative comment, which I think slightly unfair, is that Torrance did not much engage with the "veiled elusiveness of the quantum world" (see Torrance, 1985a, p. 42). He is notably much more dismissive of Pannenberg. Marley, 1992, p. 19, notes Moltmann's indebtedness to Torrance in a letter by the former dated October 6th 1986.

[48] Colyer, 2001a, p. 41.

Torrance acquits himself well, certainly better than Feyerabend, though there are weaknesses in Torrance's arguments which I have outlined. Furthermore there is a, perhaps unavoidable, trade-off between general validity and theological integrity, and Torrance positions himself at one particular end of that bargain, increasing the likelihood that whatever its merits, his work on scientific methodology is likely to remain – as we have noted it is – neglected by the world of natural science.

BIBLIOGRAPHY

Abelard, P., *Collationes*. ed. & trans. J. Marenbon & G. Orlandi, OUP, Oxford, 2001.

Achinstein, P., *Concepts of science, a philosophical analysis*. John Hopkins Press, Baltimore and London, 1968.

Achtemeier, P. M., "*The truth of tradition: Critical realism in the Thought of Aladair MacIntyre and T. F. Torrance*" in *Scottish Journal of Theology*, Vol. 47, No 3, 1994, p. 355–74.

Ackrill, J. L., *Aristotle the Philosopher*. OUP, Oxford, 1981.

Agassi, J., "*Review of Against Method*", in *Philosophia*. 6, 1976.

Alston, W., *Perceiving God: The Epistemology of Religious Experience*. Cornell University Press, Ithaca and London, 1991.

Anscombe, G. E. M., *An introduction to Wittgenstein's Tractatus*. Thoemes Press, Bristol, 1996.

Anselm, *Cur Deus Homo*. John Grant Press, Edinburgh, 1909.

Anselm, *Proslogian, with replies of Gaunilo & Anselm, Hackett*, Indianapolis & Cambridge, 2001.

Anselm, *Three philosophical dialogues*. Hackett, Indianapolis & Cambridge, 2002.

Appelyard, B., *Understanding the present science and soul of modern man*. Doubleday, New York, 1994.

Aquinas, *Summa Theologica – part 1, from Basic writings of St Thomas Aquinas ed, annotated and with intro. by Anton Pegis, Vol. I & II*. Random House, 1943–35.

Aquinas, *Summa contra gentiles, from Basic writings of St Thomas Aquinas ed, annotated and with intro. by Anton Pegis. Vol. I & II*, Random House, 1943–35.

Aquinus, *Summa Theologica II-II, from Basic writings of St Thomas Aquinas ed, annotated and with intro. by Anton Pegis, Vol. I & II*. Random House House, New York, 1943–45.

Archer, Bhaskar, Collier, Lawson, Norrie (eds.), *Critical realism: essential readings*. Routledge, London & New York, 1998.

Aristotle, *Metaphysics Γ Δ Έ, translated with notes by Christopher Kirwan*. Clarendon Press, Oxford, 1971.

Aristotle, *Metaphysics M & N, translated with introduction and notes by Julia Annas*. Clarendon Press, Oxford, 1976.

Aristotle, *Metaphysics, A, α, B, Z, H, Θ, I, K, & Λ, from the works of Aristotle*. trans. Ross W. D., Clarendon Press, Oxford, 1928.

Aristotle, *Physics I, II, translated with introduction and notes by W. Charlton*. Clarendon Press, Oxford, 1976.

Aristotle, *Posterior Analytics, translated with a commentary by Jonathan Barnes*. Clarendon Press, Oxford, 1994.

Athanasius, *Against the heathen, Nicene and Post Nicene Fathers, 2nd Series, Vol. 4*. T & T Clark, 1991.

Athanasius, *De Decretis, Nicene and Post Nicene Fathers, 2nd Series, Vol. 4*. T & T Clark, 1991.

Athanasius, *De Synodis, Nicene and Post Nicene Fathers, 2nd Series, Vol. 4*. T & T Clark, 1991.

Athanasius, *Expositio Fidei, Nicene and Post Nicene Fathers, 2nd Series. Vol. 4*, T & T Clark, 1991.

Athanasius, *On the incarnation*. Eastern Orthodox Books, (no date or city of publication cited).

Avis, P. D. L., "*Apologist from the world of science: John Polkinghorne FRS*", in *Scottish Journal of Theology*. Vol. 43, No 4, 1990, p. 485–502.
Avis, P., "*Friedrich Schleiermacher and the Science of Theology*" in *Scottish Journal of Theology*, Vol. 32, 1979, p. 19.
Bacon, F., *Novum Organum: with other parts of the Great Instauration*. translated and edited by Peter Urbach and John Gibson, Open Court, Chicago, 1994.
Barth, K., Anselm: *Fides Quaerens Intellectum*. SCM Press, London, 1930.
Barth, K., *Church Dogmatics, The Doctrine of God, 2.2*. T & T Clark Edinburgh, trans. T. F. Torrance & , G. W. Bromiley, 1956.
Barth, K., *Dogmatics in Outline*. SCM Press, London, 1949.
Barth, K., *Protestant Theology in the Nineteenth Century*. SCM Press, London 1972.
Barth, K., *The Church Dogmatics, The Doctrine of the Word of God, 1.1*. T & T Clark Edinburgh, trans. T. F. Torrance & , G. W. Bromiley, 1956.
Barth, K., *The Church Dogmatics, The Doctrine of the Word of God 1.2*. T & T Clark Edinburgh, trans. T. F. Torrance & , G. W. Bromiley, 1956.
Barth, K., *The Church Dogmatics, The Doctrine of creation, 3.2*. T & T Clark Edinburgh, trans. T. F. Torrance & , G. W. Bromiley, 1960, pages 79–95.
Barth, K., *The Church Dogmatics, The Doctrine of God, 2.1*. trans. G. W. Bromiley, T. F. Torrance, T&T Clark, Edinburgh, 1957.
Barth, K., *The Epistle to the Romans*. Oxford University Press, London, Oxford, New York, 1933.
Barth, K., *The Faith of the Church*. Fontana Books, 1969.
Barth, K., *The Knowledge of God and the service of God, according to the teaching of the Reformation (Gifford lectures 1937–8)*. Hodden and Stoughton, London, 1938.
Barth, K., *The Word of God and the Word of Man*. Peter Smith, 1928.
Bartholomew, D. J., *God of Chance*. SCM Press, London, 1984.
Bauer, W., *Orthodoxy and Heresy in earliest Christianity*. eds. R. Kraft & G. Krodel, Fortress Press, Philadelphia, 1979.
Beattie, T., *The New Atheists*. Darton, Longman & Todd, London, 2007.
Berger, P. L., & T. Luckmann, *The Social Construction of Reality, A Treatise in the Sociology of Knowledge*. Penguin Books, London, 1971.
Bernstein, R. J., *Beyond Objectivism and Relativism*. Blackwell, Oxford, 1983.
Bhaskar, R., "*Feyerabend and Bachelard: Two Philosophies of Science*" in *New Left Review*, 94, 1975.
Bhaskar, R., *Philosophy and the idea of Freedom*. Blackwells, Oxford 1991.
Bhaskar, R., *The Possibility of Naturalism, 3rd edition*. Routledge, London and New York, 1998
Brown, J., *Subject and Object in Modern Theology*. SCM Press, London, 1955.
Brunner, E., (*Nature and Grace*.), and K. Barth (*NO!*), *Natural theology*. trans. Fraenkel, P., Intro. by Baillie, J., Geoffrey Bles: The Centenary Press, London, 1946.
Brunner, E., *Revelation and Reason*. SCM Press, London, 1947.
Buber, M., *Eclipse of God, Studies in the Relation between Religion and Philosophy*. Harpers & Brothers, New York, 1957.
Buber, M., *I and Thou*. T&T Clark, Edinburgh 1970.
Bultmann, R., *Jesus and the Word*. Fontana books, 1958.
Bultmann, R., *New Testament and Mythology and other writings*. SCM, London, 1984.
Burian, R. M., "*Scientific Realism and Incommensurability: Some Criticisms of Kuhn and Feyerabend*", *in R.S. Cohen and M.W. Wartofsky Methodology, Metaphysics and the History of Science*. eds., Reidel, Dordrecht, 1984.
Burridge, R., *What are the Gospels?*, CUP, Cambridge, 1992.
Calvin, J., *Institutes of the Christian Religion, Vol. 1 & 2*. J. T. McNeill (ed.), trans. F. L. Battles, Westminster Press, Philadelphia,1960/1961.
Carvin, W. P., "*Creation and Scientific Explanation*" in *Scottish Journal of Theology*, Vol. 36, 1983, p. 289.

Charon, J., *Cosmology*. translated by Patrick Moore, Weidenfeld and Nicolson, London, 1970.

Chomsky, N., *Language and Mind*. Harcourt, Brace and World, New York, 1968.

Clark, S. R. L., "*Feyerabend's Conquest of Abundance*" in *Inquiry* 45, 2002, p. 249–68.

Colyer, E., ed. Lanham, *The promise of Trinitarian theology: Theologians in Dialogue with T. F. Torrance*. Rowman & Littlefield, 2001.

Colyer, E., *How to Read T. F. Torrance: Understanding his Trinitarian and Scientific Theology*. Intervarsity Press, Illinois, 2001.

Connolly, J., & Keutner, T., (eds.), *Hermeneutics versus Science? Three German Views: Wolfgang Stegmuller, Hand-Georg Gadamer, Ernst Konrad Specht*. University of Notre Dame Press, Indiana, 1988.

Couvalis, G., *Feyerabend's Critique of foundationalism*. Avebury, Sydney, 1989.

Crombie, I. M., *Plato – The Midwife's Apprentice*. Routledge & Kegan Paul, London, 1964.

Cupitt, D., *Taking leave of God*. SCM Press, London, 1980.

Cupitt, D., *The Sea of faith: Christianity in Change*. BBC, London, 1985.

Descartes, R., "*Discourse on the Method & Meditations on First Philosophy*", in *The Essential Descartes*, p. 106–223. Wilson, M. (ed.), Meridian, New York, 1983.

Devitt, M., "Against Incommensurability" in *Australasian Journal of Philosophy*, 57, 1975.

Diem, H., *Dogmatics*. trans. H. Knight, Oliver and Boyd, Edinburgh, 1959.

Dilthey, W., *Selected writings*. ed., trans., and intro. by H. P. Rickman, CUP, London, New York, Melbourne, 1976.

Dreyfus, H., & Rabinow, P., *Michel Foucault, Beyond Structuralism and Hermeneutics*. University of Chicago Press, Chicago, 1982.

Duhem, P., *The Aim and Structure of Physical Theory*, translated from the second French edition by P. P. Wiener, Princeton UP, Oxford UP, 1954.

Dummett, M., *Truth and other enigmas*. Harvard University Press, Cambridge, Massachusetts, 1978.

Eddington, A., *The Philosophy of Physical Science*. Cambridge University Press, 1949.

Einstein, A., *Essays in Science*. Trans. from the German by Harris A., New York Philosophical Library, New York, 1934.

Einstein, A., *Relativity*. Routledge, London & New York, 1993.

Evans, G. R., *Anselm*. Geoffrey Chapman, London, 1989.

Evans-Pritchard, E. E., *Theories of Primitive Religion*. Clarendon Press, Oxford, 1965.

Farell, R., "Feyerabend's Metaphysics: Process-Realism, or Voluntarist-Idealism?" in *Journal for General Philosophy of Science*, Vol. 32/2, 2001, p. 351-69, Springer, 2001.

Farrer, A., *The Glass of Vision (Brampton Lectures 1948)*. Dacre, London 1948.

Fergusson, D., (ed.), *Scottish Philosophical Theology 1700–2000*. Imprint Academic, Exeter & Charlottesville, 2007.

Feyerabend, P., & Lakatos, I., *For and against method*. Ed. & intro. M. Motterlini, University of Chicago Press, Chicago, 1999.

Feyerabend, P., *Against Method*. 3rd edition, Verso, London, New York, 1993 (earlier editions 1975, 1988).

Feyerabend, P., *Conquest of Abundance*. ed. Terpstra, B., University of Chicago Press, Chicago & London, 1999.

Feyerabend, P., *Farewell to Reason*. Verso, London & New York 1987.

Feyerabend, P., *Killing Time*. University of Chicago Press, Chicago, 1995.

Feyerabend, P., *Knowledge without Foundations, lectures in Oberlin College*. Ohio, 1961.

Feyerabend, P., *Knowledge, Science and Relativism*. Ed. J. Preston, Cambridge University Press, Cambridge, 1999.

Feyerabend, P., *Problems of empiricism, Philosophical Papers, Vol. 2*. Cambridge University Press, 1981.

Feyerabend, P., *Realism, rationalism and the scientific method, Philosophical Papers, Vol. 1*. Cambridge University Press, 1981.
Feyerabend, P., "*Science Freedom and Good Life*" in *Philosophical Journal*, 1968.
Feyerabend, P., *Science in a free society*. London NLB, London, 1978.
Feyerabend, P., *Three Dialogues on Knowledge*. Oxford Cambridge, Mass: Blackwell, 1991.
Finocchiaro, M. A., "*Rhetoric and Scientific Rationality*" in P. D. Asquith & I. Hacking (eds.), *PSA 1978*, Volume 1, 1978.
Flew, A., *God and Philosophy*. Prometheus Books, New York, 2005.
Floyd, J., "Homage to Vienna: Feyerabend on Wittgenstein (and Austin and Quine)" in K. R. Fischer & F. Stadler (eds.), *Paul Feyerabend (1924–1994) — Ein Philosoph aus Wien*, Springer Verlag, 2005.
Floyd, J., "Review of P.K. Feyerabend, Against Method, and Farewell to Reason" in *Journal of Philosophy*, 88, 1991.
Ford, D., with R. Muers, (eds.), *The Modern Theologians*. Blackwell Publishing, Malden, Oxford, Victoria, 2005.
Frege, G., *The Frege Reader*. Ed. Michael Beaney, Blackwell, Oxford, 1997.
Frei, H. W., *The Eclipse of Biblical Narrative: A study in Eighteenth and Nineteenth Century Hermeneutics*. Yale University Press, New Haven and London, 1974.
Frei, H. W., *The Identity of Jesus Christ, The hermeneutical bases of Dogmatic Theology*. Fortress Press, Philadelphia, 1975.
Frei, H. W., *Types of Christian Theology*. Yale University Press, New Haven and London, 1992.
Gadamer, H. G., *Truth and Method*. Sheed & Ward Stagbooks, London, 1975.
Geldhof, J., *Revelation, reason and reality: theological encounters with Jaspers, Schelling and Baader*. Peeters, Leuven, Paris, Dudley, 2007.
Gelwick, R., *The Way of Discovery*. Oxford University Press, New York, 1977.
Godel, K., *Collected Works, Volumes 1 & 2*. ed. Solomon Feferman, Oxford University Press, Oxford, 1990.
Griffith-Dickson, G., "'Outsidelessness' and the 'beyond' of signification" in *Heythrop Journal*, 1996b, p. 258–73.
Griffith-Dickson, G., *The Philosophy of Religion*. SCM, London, 2005.
Griffith-Dickson, G., *Human and Divine*. Duckworth, London, 2000.
Griffith-Dickson, G., *Johann Georg Hamann's Relational Metacriticism*. De Gruyter, Berlin, New York, 1995.
Griffith-Dickson, G., *Teaching philosophy of religion 'multiculturally': a Lokahi approach?* Unpublished paper.
Grunfeld, J., "Feyerabend's Flight from Reason" in *Science et Esprit*, Vol. 42, 1990, p. 89–101.
Grunfeld, J., "Method and Language" in *Science et Espirit*, Vol. 30, 1978, p. 185–192.
Grunfeld J., "Rationality and scientific method" in *Science et Esprit*, Vol. 28 1976, p. 309–21.
Habermas, J., *Knowledge and Human Interests*. Boston, Mass. and London, 1972.
Hacking, I., *The Social construction of what?* Harvard University Press, Cambridge and London, 2000.
Hacking, I., *Why does language matter to philosophy?* Cambridge University Press, 1975.
Hare, R. M., *Plato*. Oxford University Press, Oxford, 1982.
Harré, R., "Review of P.K. Feyerabend's Against Method" in *Mind*, 86, 1977.
Harris, E. E., *Revelation through Reason*. Allen & Unwin, London 1959.
Heidegger, M., *Being and Time*. SUNY Press, Albany & New York, 1996.
Heim, K., *The World: Its Creation and Consummation*. Trans. R. Smith, Muhlenberg Press, Philadelphia, 1962.
Heisenberg, W., *Physics and philosophy*. George Allen and Unwin, London, 1959.

Helm, P., *The Divine Revelation*. Marshall, Morgan & Scott, London, 1982.
Hentschel, K., "On Feyerabend's Version of Mach's Theory of Research and its Relation to Einstein". *Studies in History and Philosophy of Science*, 16, 1985.
Hing-Kau Yeung, J., *Being and Knowing, An Examination of T. F. Torrance's Christological Science*. Alliance Bible Seminary, Hong Kong, 1996.
Hodges, W., *Logic*. Penguin Books, Harmondsworth, New York, Victoria, Ontario, Aukland, 1977.
Honderich, T., "On Benjamin Libet: Is the mind ahead of the brain? Or behind it?" in: *On determinism and Freedom*, Edinburgh University Press,
Honderich, T., "The time of a conscious sensory experience and mind-brain theories" in: *Journal of Theoretical Biology*, Vol. 110, 1984, p. 115–29.
Hume, D., *A Treatise of Human Nature*. Ed. L. A. Selby-Bigge, Clarendon, Oxford, 1978.
Hume, D., *Theory of Knowledge*. Ed. D. C. Yalden-Thomson, The Nelson Philosophical Texts, 1951.
Husserl, E., *Cartesian Meditations: An Introduction to Phenomenology*. Trans. D. Cairns, Nijhoff, The Hague, 1960.
Husserl, E., *Phenomenology and the Crisis of Philosophy as Rigorous Science,: and Philosophy and the Crisis of European Man*. Trans. with notes and introduction by Q. Lauer, Harper & Row, New York & London, 1965.
Jaki, S. L., "Science and Christian Theism: A Mutual Witness" in: *Scottish Journal of Theology*, Vol. 32, 1979, p. 563.
Jammer, M., *Einstein and Religion*. Princeton University Press, 1999.
Jeanrond, W., *Theological Hermeneutics: Development and Significance*. SCM, London, 1994.
John Paul II, *Faith and Reason, Encyclical letter*. The Catholic Truth Society, London, 1998.
Jones, W. B., "Theory-Ladenness and Theory Comparison" in: P.D.Asquith & I.Hacking (eds.), *PSA* 1978, Volume 1. East Lansing, Mi: Philosophy of Science Association, 1975.
Kant, I., *Critique of Pure Reason*. Cambridge University Press, 1998.
Kant, I., *On the Miscarriage of all Philosophical Trials in Theodicy*. Cambridge University Press, 1998.
Kant, I., *Religion within the boundaries of mere reason*. Cambridge University Press, 1998.
Kant, I., *The conflict of the faculties*. Trans. M. J. Gregor, University of Nebraska Press, Lincoln & London, 1979.
Kant, I., *The end of all things*. Cambridge University Press, 1998.
Kant, I., *What does it mean to orient oneself in thinking?* Cambridge University Press, 1998.
Kant I., "*What is Enlightenment?*" in *Selections*, 1996, p. 142.
Kerr, F., *Theology after Wittgenstein*. SPCK, London, 1997.
Khang, P. S., "The Epistemological Significance of Homoousion in the theology of Thomas F. Torrance" in *Scottish Journal of Theology*, Vol. 45, 1992, p. 341.
Kierkegaard, S., *Concluding Unscientific Postscript to Philosophical Fragments, Vol. 1 & 2*. ed. & trans. H. Hong and E. Hong, Princeton UP, Princeton NJ, 1992.
Kierkegaard, S., *Philosophical Fragments*. Originally trans. D. F. Swendon, new intro. and commentary N. Thulstrup, trans. revised and commentary trans. H. Hong, Princeton UP, Princeton NJ, 1946.
Kirby, R. S., *The Theological Definition of Cosmic Disorder in the Writings of Thomas Forsyth Torrance*. Unpublished PhDthesis, Kings College London, 1992.
Kirk, G. S., J. E. Raven, & M. Scholfield, *The Presocratic Philosophers: a Critical History with a Selection of Texts*. Cambridge UP, Cambridge, 1962.
Kirk, R., *Relativism and Reality*. Routledge, London & New York, 1999.

Klinefelter, D. S., "God and Rationality: A Critique of the Theology of Thomas F. Torrance" in: *The Journal of Religion*, Vol. 53, No 1, 1973, p. 117–35.
Kruger, C. B., "The Doctrine of the knowledge of God in the Theology of T. F. Torrance: Sharing in the Son's communication with the Father in the Spirit" in *Scottish Journal of Theology*, Vol. 43, 1990, p. 366.
Küng, H., and D. Tracy, (eds.), *Paradigm Change in Theology*. T & T Clark, Edinburgh, 1989.
Kuhn, T. S., *The Structure of Scientific Revolutions*. University of Chicago Press, Chicago, 1970.
Lakatos, I., *Proofs and Refutations: the Logic of Mathematical Discovery*. Ed. J. Worrall & E. Zahar, Cambridge UP, Cambridge, 1976.
Lakatos, I., *The Methodology of Scientific Research Programmes*. Ed. J. Worrall & G. Currie, Cambridge University Press, Cambridge, 1978.
Lang Uwe, *John Philoponus and the Controversies over Chalcedon in the Sixth Century: A study and translation of the Arbiter*. Peeters, Leuvan, 2001.
Langford, T. A., "T. F. Torrance's Theological Science: a reaction" in *Scottish Journal of Theology*, Vol. 25, 1972, p. 155–70.
Laudan, L., *Beyond Positivism and Relativism*. Westview Press, Colorado, 1996.
Laudan, L., *Progress and Its Problems; Towards a Theory of Scientific Growth*. University of California Press, Los Angeles & London, 1977.
Lee, N., *Calvin on the Sciences*. Burlington Press, Cambridge, 1969.
Lindbeck, G., *The Nature of Doctrine*. SPCK, London, 1974.
Luoma, T., *Incarnation and Physics: Natural science in the theology of Thomas F. Torrance*. OUP, Oxford 2002.
MacIntyre, A., *Whose justice? Which rationality?* Duckworth Press, London, 1988.
MacKay, D. M., *Science, Chance, & Providence*. OUP, Oxford, 1978.
MacMurray, J., *Search for Reality in Religion*. Quaker Home Service, London, 1995.
MacMurray, J., *The Personal World, on Self and Society*., sel. & ed. P. Conford, Floris Books, Edinburgh, 1996.
MacMurray, J., *The Philosophy of Jesus*. Friends Home Service Committee, London, 1973.
Marley, A., *T. F. Torrance: The Rejection of Dualism*. Handsel, Edinburgh, 1992.
Maxwell, J., *A Dynamical Theory of the Electromagnetic Field*. Intro. and ed. T. F. Torrance, appreciation by A. Einstein, Scottish Academic Press, Edinburgh, 1982.
McCabe, H., *Faith within Reason*, Continuum books, London & New York, 2007.
McDonald, N. B., *Karl Barth and the Strange New World within the Bible*, Paternoster Press, Carlisle, 2000.
McDowell, J. C., & M. Higton, (eds.), *Conversing with Barth*. Ashgate, Aldershot UK, Burlington USA, 2004.
McGrath, A., *T. F. Torrance, An Intellectual Biography*. T & T Clarke, Edinburgh, 1999.
McGrath, A., *A scientific theology, Vol. 1 – Nature*. T&T Clarke, London and New York, 2001.
McGrath, A., *A scientific theology, Vol. 2 - Reality*. T&T Clarke, London and New York, 2002.
McGrath, A., *A scientific theology, Vol. 3 – Theory*. T&T Clarke, London and New York, 2003.
McGrath, A., *The Foundations of Dialogue in Science and Religion*. Blackwells, Oxford and Massachusetts, 1998.
McGrath, A., *The Order of Things: Explorations in Scientific Theology*. Blackwell, Oxford, 2006.
McGrath, A., *The science of God*. T & T Clark, London & New York, 2004.
McKinney, R. W. A., *Creation Christ and Culture, Studies in honour of T. F. Torrance*. T & T Clark, Edinburgh, 1976.

Mill, J. S., *On Liberty*. Penguin Classics, 1974.
Moberg, D. W., *Are there Rival, Incommensurable Theories?* Philosophy of Science, 46, 1979.
Molnar, P. D., "God's self communication in Christ: A comparison of Thomas F. Torrance and Karl Rahner" in *Scottish Journal of Theology*, Vol. 50, 1997, p. 288.
Moore, A., & M. Scott, (eds.), *Realism and Religion*. Ashgate, Aldershot, Burlington, 2007.
Moore, A., *Realism and Christian faith: God, grammar and meaning*. Cambridge UP, Cambridge, 2003.
Morgan, R., with J. Barton, *Biblical Interpretation*. OUP, Oxford, 1988.
Morrison, J. D., *Knowledge of the Self-revealing God in the Thought of Thomas Forsyth Torrance*. Peter Lang, New York, 1997.
Murphy, N., *Theology in the Age of Scientific Reasoning*. Cornell University Press, Ithaca and London, 1990.
Murtha-Smith. S., "The Advancement of New Theology using New Science: the Three Key Concepts of Thomas Torrance" in *Journal of Faith and Science Exchange*, 1997, p. 65–71.
Nebelsick, H. P., "Iain Paul: Science, Theology and Einstein" in: *Scottish Journal of theology*, Vol. 37, 1984, p. 237.
Need, S. W., *Truly Divine and Truly Human: The Story of Christ and the Seven Ecumenical Councils*. SPCK, London, Hendrickson, Peabody, 2007.
Neidhardt, W., "Thomas F. Torrance's Integration of Judeo-Christian Theology and Natural Science: Some Key Themes" in: *Perspectives on Science and Christian faith*, Vol. 41, No 2, 1989, p. 87–90.
Newton-Smith, W. H., *The Rationality of Science*. Routledge, London New York, 1990.
Niebuhr, H. R., *Christ and Culture*. Harper & Row, New York, 1956.
Nielsen, K. & D. Z. Phillips, *Wittgenstein Fideism*. SCM Press, London, 2005.
Norris, F., "Mathematics, Physics and Religion: a Need for Candor and Rigor" in: *Scottish Journal of Theology*, Vol. 37, No 4, 1984, p. 457–70.
Oberdan, T., "The Vienna Circle's 'anti-foundationalism'" in: *The British Journal for the Philosophy of Science*, Vol. 49/2, p. 297–308, Oxford University Press, 1998.
Ockham, W., *Philosophical Writings*. Ed. & trans. Boehner, P., Nelson Philosophical texts, London, 1957.
Origen, *De Principiis from The Anti-Nicene Fathers, Vol. 4*. T & T Clark, 1994.
Osborne, G. R., *The Hermeneutical Spiral: A Comprehensive Introduction to Biblical Interpretation*. IVP Academic, Illinois, 2006.
Palmer, R. E., *Hermeneutics: Interpretation theory in Schleiermacher, Dilthey, Heidegger and Gadamer*. Northwestern University Press, Evanston, 1969.
Pannenberg, W., *Jesus: God and Man*. Trans. Wilkins, L., & Priebe, D., SCM Press, London, 1968.
Pannenberg, W., "Problems between Science and Theology in the Course of their Modern History" in: *Zygon*, Vol. 41, No 1, March 2006, p. 105–12 (8), Blackwell Publishing, 2006.
Pannenberg, W., *Theology and the Philosophy of Science*. Trans. McDonagh, F., Darton, Longmann and Todd, London, 1976.
Patterson, S., *Realist Christian Theology in a Postmodern Age*. Cambridge University Press, 1999.
Paul, I., *Science Theology and Einstein*. Christian Journals, Belfast, 1982.
Peacocke, A., *Paths from Science towards God; the End of all our Exploring*. Oneworld, Oxford, 2001.
Peacocke, A., *Theology for a Scientific Age*. SCM, London, 1993.
Penrose, R., *The Emperor's New Mind*. Oxford UP, Oxford, 1989.
Philoponus, John – *On Aristotle on coming to be and perishing, vol. 1 & 2*. Trans. C. J. F. Williams, intro. S. Berryman, Duckworth Press, London, 1999.
Phillips, D. Z., *Belief, Change and Forms of Life*. MacMillan, London, 1986.

Phillips, D. Z., *Religion and the Hermeneutics of Contemplation*. CUP, Cambridge, 2001.
Planck, M., *The Philosophy of Physics*. Trans. W. H. Johnson, Allen & Unwin, London, 1936.
Plantinga, A., & M. Tooley, *Knowledge of God*. Blackwell, Malden, Oxford, Victoria, 2008.
Plantinga, A., & N. Wolterstorff, (eds.), *Faith and Rationality: Reason and Belief in God*. University of Notre Dame Press, Indiana, 1983.
Plantinga, A., *The Analytic Theist, An Alvin Plantinga reader*. Eds. Sennett, J. F., William B. Eerdmans Publishing, Grand Rapids Michigan, Cambridge UK, 1998.
Plato, "*Phaedo*., in: *Plato Complete Works*, ed. J. M. Cooper, Hackett, Indianapolis, 1997, p. 49ff.
Plato, "*Theatetus*" in *Complete Works*, ed. J. M. Cooper, Hackett, Indianapolis, 1997, p. 179ff.
Plato, *Timaeus*. Trans. & intro. H. D. P. Lee, Penguin Books, Baltimore, 1965.
Platt ,T., "The Conflict of Science and Religion: a Confusion Re-visited" in: *Journal of Interdisciplinary Studies*, Vol. 2, p. 1990, p. 22–38.
Polanyi, M., & Prosch, H., *Meaning*. University of Chicago, Chicago, 1977.
Polanyi, M., *Knowing and Being*. Ed. M. Green, Routledge & Kegan Paul, London, 1969.
Polanyi, M., *Personal Knowledge*. Routledge, London, 1998.
Polanyi, M., *Science, Faith and Society*. University of Chicago Press, Chicago, 1946.
Polanyi, M., *The Logic of Personal Knowledge: Essays presented to M. Polanyi on his Seventieth Birthday*. Routledge & Kegan Paul, London 1961.
Polanyi, M., *The Tacit Dimension*. Routledge & Kegan Paul, London, 1967.
Polkinghorne, J., *Science and Christian Belief*. SPCK, London, 1994.
Polkinghorne, J., *Reason and Reality: The Relationship between Science and Theology*. SPCK, London 1991.
Polkinghorne, J., *Science and Theology*. Westview Press, Colorado, 1998.
Polkinghorne, J., *Belief in God in an Age of Science*. Yale University Press, New Haven & London, 1998.
Polkinghorne, J., *Faith, Science and Understanding*. SPCK, London, 2000.
Polkinghorne, J., *Quantum Physics and Theology: An Unexpected Kinship*. SPCK, London, 1997.
Polya, G., *How to Solve It: a New Aspect of Mathematical Method*. Princeton UP, Princeton, 1971.
Popper, K., *Conjectures and Refutations: the Growth of Scientific Knowledge*. Routledge & Paul Kegan, London, 1969.
Popper, K., *Objective Knowledge: an Evolutionary Approach*. Clarendon Press, Oxford, 1979.
Popper, K., *Realism and the Aim of Science*. Hutchinson, London et al, 1983.
Popper, K., *The Logic of Scientific Discovery*. Routledge, London & New York, 1992.
Preston, J., *Feyerabend, Philosophy Science & Society*. Polity, Cambridge, 1997.
Preston, J., G. Munévar, D. Lamb, (eds.), *The Worst Enemy of Science? Essays in Memory of Paul Feyerabend*. Oxford University Press, New York & Oxford, 2000.
Preston, J., "Science as Supermarket: "Post-Modern" Themes in Paul Feyerabend's Later Philosophy of Science" in: *Studies in History and Philosophy of Science*, 29, 1998.
Putnam, H., *The Many Faces of Realism*. La Salle, Open Court, 1987.
Quine, W. V., *"Philosophy of Logic", in Truth*. Blackburn, S., & K. Simmons, (eds.), Oxford University Press, Oxford 1999.
Quine, W. V., *Pursuit of Truth*. Harvard University Press, Cambridge Massachusetts, and London, 1990.
Quine, W. V., *The Philosophy of W V Quine*. L. E. Hahn & P. A. Schlipp (eds.), expanded edition 1998, Open Court, Chicago & Illinois, 1998.

Quine, W. V., *Word and Object*. MIT Press, Cambridge, Massachusetts, 1960.
Rae, A., *Quantum Physics: Illusion or Reality?* CUP, Cambridge, 2004.
Rae, M. A., *History and Hermeneutics*. T & T Clark, London and New York, 2005.
Reichenbach, H., *Modern Philosophy of Science*. Routledge and Keegan Paul, London, 1959.
Ricoeur, P., *Essays on Biblical Interpretation*. Intro. & ed. L. S. Mudge, SPCK London, 1981.
Ritchie, A. D., *Essays in Philosophy*. Longmans, Green & Co, London, New York, Calcutta, 1948.
Rorty, R. (et al.), eds. Brandom and Malden, *Rorty and his Critics*. Blackwell, Massachusetts & Oxford, 2000.
Rorty, R., *Objectivity, Relativism and Truth (Philosophical Papers)*. Cambridge UP, Cambridge, 1991.
Rousseau, J. J., *The Discourses*. Ed. & trans. V. Gourevitch, CUP, Cambridge & New York, 1997.
Rupp, E., & B. Drewery, (ed.), *Martin Luther*. Edward Arnold, London, 1970.
Russell, B., *History of Western Philosophy*. Routledge, London, 1996.
Sankey, H., *Semantic Incommensurability and Scientific Realism*. Unpublished paper, 2007.
Scheffler, I., *Science and Subjectivity*. Hackett, Indianopolis, 1982.
Schiff, L., *Quantum Mechanics, Third International Student edition*. McGraw and Hill, Auckland et al, 1968.
Schleiermacher, F., *The Christian Faith*. T & T Clark, Edinburgh, 1928.
Schmidt, L. K., *Understanding Hermeneutics*. Acumen, Stocksfield, 2006.
Schubert, F., "Thomas F. Torrance: The Case for a Theological Science" in *Encounter*, Vol. 45 1984, p. 123–37.
Scott, D., *Micheal Polanyi*. (2nd preface by Torrance), SPCK, London, 1995.
Seimens, D., "Two problems with Torrance" in: *Perspectives on Science and Christian faith*, Vol. 43, No2, 1991, p. 112–3.
Shea, W., & M. Artigas, *Galileo in Rome*. OUP, Oxford, 2003.
Shults, F., "A theology of Chaos: An Experiment in Postmodern Theological Science" in *Scottish Journal of Theology*, Vol. 45, 1992, p. 223.
Smedes, T., *Chaos, Complexity and God: Divine Action and Scientism*. Peeters, Leuven, Paris, Dudley, 2004.
Smith, R., J. G. Hamaan, *A study in Christian Existence*. (includes a selection of writings of Hamaan), Collins, London, 1960.
Sobel, D., *Galileo's Daughter*. Fourth Estate, London, 1999.
Sorabji, R., (ed.), *Philoponus and the Rejection of Aristotelian Science*. Duckworth, London, 1987.
Soskice, J. M., *Metaphor and Religious Language*. Clarendon Press, Oxford, 1985.
Southgate, C., (ed.), *God, Humanity and the Cosmos : A textbook in Science and Religion*. Continuum T & T Clark, London & New York, 1999.
Swinburne, R., *Epistemic Justification*. Clarendon Press, Oxford, 2001.
Swinburne, R., ed., *The Justification of Induction*. OUP, Oxford, 1974.
Swinburne, R., *Faith and Reason* OUP, Oxford, 1981.
Swinburne, R., *Revelation*. Clarendon Press, Oxford, 2004.
Swinburne, R., *The Evolution of the Soul*. Clarendon, Oxford, 1986.
Swinburne, R., *The Existence of God*. Clarendon Press, Oxford, 1991[1].
Tambiah, S., *Magic, Science, Religion and the Scope of Rationality*. CUP, Cambridge, 1990.

[1] I read this book before the publication of the most recent 2004 edition. References in the text refer to this 1991 edition. With reference to this book I am not aware that the latest edition has any significant implications.

Tarski, A., *Introduction to Logic, and to the Methodology of Deductive Sciences*. OUP, New York London, 1965.

Tarski, A., *Logic, Semantics, Metamathematics – Papers 1923–38*. Trans. J. H. Woodger, Hackett, Indianapolis, 1938.

Taylor, A., *The Faith of a Moralist - Volume II, Natural Theology and the Positive Religions, Gifford Lectures 1927–8*. MacMillan and Co, London, 1930.

Thiselton, A. C., *New Horizons in Hermeneutics: The Theory and Practice of Transforming Biblical Reading*. Harper Collins, Glasgow, 1992.

Torrance, T.F., (ed.), *Belief in Science and in Christian Life: The Relevance of Michael Polanyi's Thought for Christian Faith and Life*. Handsel Press, Edinburgh, 1980.

Torrance, T.F., *Calvin's Doctrine of Man*. Lutterworth Press, London, 1949.

Torrance, T.F., *Christian Theology and Scientific Culture*. Christian Journals, Belfast, 1980.

Torrance, T.F., *Divine and Contingent Order*. T & T Clark, Edinburgh 1981.

Torrance, T.F., *Divine Meaning, Studies in Patristic Hermeneutics*. T & T Clarke, Edinburgh, 1995.

Torrance, T.F., *God And Rationality*. T & T Clarke, Edinburgh, 1971.

Torrance, T.F., "Hermeneutics according to F. D. E. Schleiermacher" in *Scottish Journal of Theology*, Vol. 21, 1968, p. 257.

Torrance, T.F., "Karl Barth and the Latin Heresy" in the *Scottish Journal of Theology*, Vol. 39, 1986a, p. 461.

Torrance, T.F., *Karl Barth, Biblical and Evangelical Theologian*. T & T Clark, Edinburgh, 1990.

Torrance, T.F., *Reality and Evangelical Theology*. The Westminster Press, Philadelphia. 1982.

Torrance, T.F., *Reality and Scientific Theology*. Scottish Academic Press, Edinburgh, 1985.

Torrance, T.F., "Revelation, Creation, and Law" in *Heythrop Journal*, 1996b, p. 273–83.

Torrance, T.F., *Royal Priesthood, Scottish Journal of Theology occasional papers no. 3*. Oliver & Boyd Ltd., Edinburgh, 1955a.

Torrance, T.F., *Scottish Theology*. T & T Clark, Edinburgh, 1996a.

Torrance, T.F., *Space Time and Incarnation*. T & T Clark, Edinburgh 1969a.

Torrance, T.F., *Space Time and Resurrection*. The Handsel Press, Edinburgh, 1976.

Torrance, T.F., *The Christian Doctrine of God: One Being Three Persons*. T & T Clark, Edinburgh, 1996.

Torrance, T.F., *The Christian Frame of Mind*. Handsel Press, Edinburgh, 1985a.

Torrance, T.F., *The Doctrine of Grace in the Apostolic Fathers*. Oliver and Boyd, Edinburgh, 1948.

Torrance, T.F., *The Ground and Grammar of Theology*. T & T Clark, Edinburgh & New York, 2001.

Torrance, T.F., *The Hermeneutics of John Calvin*. Scottish Academic Press, Edinburgh, 1988a.

Torrance, T.F., (ed.), *The Incarnation: Ecumenical studies in the Nicene-Constantinople Creed AD381*. Handsel Press, Edinburgh, 1981.

Torrance, T.F., "The legacy of Karl Barth" in: T*he Scottish Journal of Theology*, Vol. 39, 1986, p. 289.

Torrance, T.F., *The Mediation of Christ*. T & T Clark, Edinburgh, 1992.

Torrance, T.F., *The Trinitarian Faith*. T & T Clark, Edinburgh,1988.

Torrance, T.F., *Theological Science*. T & T Clark, Edinburgh, 1969.

Torrance, T.F., *Theology in Reconstruction*. SCM, London, 1965.

Torrance, T.F., *Transformation and Convergence in the Frame of Knowledge*. Christian Journals Limited, Belfast, Dublin & Ottowa, 1984.

Torrance, T.F., *Trinitarian Perspectives, Toward Doctrinal Agreement*. T & T Clark, Edinburgh, 1994.
Van Fraassen, B., *The Scientific Image*. Clarendon Press, Oxford, 1980.
Van Huyssteen, J., *Essays in Postfoundationalist Theology*. William B. Eerdmans Publishing, Michigan & Cambridge, 1997.
Waismann, F., *Introduction to Mathematical Thinking: the Formation of Concepts in Modern Mathematics*. Trans. T. F. Benac, Hafner Publishing Co, 1952.
Waismann, F., *Ludwig Wittgenstein and the Vienna Circle*. Ed. B. McGuinness, trans. J. Schulte & B. McGuinness, Blackwells, Oxford, 1979.
Waismann, F., *The Principles of Linguistic Philosophy*. Ed. R. Harre, MacMillan, London, 1965.
Watson, F., (ed.), *The Open Text: New Directions for Biblical Studies*. SCM, London, 1993.
Weber, M., *The Methodology of the Social Sciences*. Trans. & ed. E. Shils & H. Finch, Free Press, New York, 1949.
Webster, J., (ed.), *The Cambridge Companion to Karl Barth*. Cambridge University Press, Cambridge, 2000.
Weightman, C., *Theology in a Polanyian Universe: The Theology of Thomas Torrance*. Peter Lang, New York, 1994.
Weizsäcker, C. von, *The World View of Physics*. Trans. Grene, M., Routledge and Kegan Paul, London, 1952.
Werth, R., "On the Theory-Dependence of Observations" in *Studies in History and Philosophy of Science*, 11, 1980.
Whorf, B. L., *Language, Thought and Reality: Selected writings of Benjamin Lee Whorf*. Ed. and intro. by J. B. Carroll, M I T Press, Masssachusetts, 1956.
Wigley, S., "Karl Barth on St Anselm: The influence of Anselm's 'Theological Scheme' on T. F. Torrance and Eberhard Jungel" in *Scottish Journal of Theology*, Vol. 46, No 1, 1993, p. 79–97.
Wittgenstein, L., *On Certainty*. Blackwell, Oxford, 2001.
Wittgenstein, L., *Philosophical Investigations*. Trans. G. E. M. Anscombe, Oxford, Blackwell, 1967.
Wittgenstein, L., *Tractatus Logico-Philosophicus*. intro. by B. Russell, trans. D. F. Pears & B. F. McGuinness, Routledge Classics, London & New York, 2001.
Worrall, J., "Feyerabend and the Facts" in: Munévar (ed.), *Beyond Reason*, Dordrecht, Kluwer, 1991.

INDEX

Studies in Systematic Theology

1. Chung, P.S. *Constructing Irregular Theology*. Bamboo and Minjung in East Asian Perspective. 2009. ISBN 978 90 04 17417 7
2. Osborne, K. *A Theology of the Church for the Third Millennium*. A Franciscan Approach. 2009. ISBN 978 90 04 17657 7
3. Huang, P. *Confronting Confucian Understandings of the Christian Doctrine of Salvation*. A Systematic Theological Analysis of the Basic Problems in the Confucian-Christian Dialogue. 2009. ISBN 978 90 04 17726 0
4. Küster, V. *A Protestant Theology of Passion*. Korean Minjung Theology Revisited. 2010. ISBN 978 90 04 17523 5
5. Tulud Cruz, G. *An Intercultural Theology of Migration*. Pilgrims in the Wilderness. 2010. ISBN 978 90 04 18122 9
6. Trice, M.R. *Encountering Cruelty: The Fracture of the Human Heart*. 2011. ISBN 978 90 04 20166 8
7. Munchin, D. *Is Theology a Science?* The Nature of the Scientific Enterprise in the Scientific Theology of Thomas Forsyth Torrance and the Anarchic Epistemology of Paul Feyerabend. 2011. ISBN 978 90 04 19459 5

Printed in the United States
By Bookmasters